STUDIES IN PHILOSOPHY

XV

STRICT FINITISM

AN EXAMINATION OF LUDWIG WITTGENSTEIN'S *REMARKS ON THE FOUNDATIONS OF MATHEMATICS*

by

CHARLES F. KIELKOPF
Ohio State University

1970
MOUTON
THE HAGUE · PARIS

LIBRARY OF CONGRESS CATALOG CARD NUMBER: 70-106464

Printed in The Netherlands by Mouton & Co., Printers, The Hague.

PREFACE

In an essay entitled "Wittgenstein's Remarks on the Foundation of Mathematics", in the *British Journal for the Philosophy of Science*, IX, No. 34 (August, 1958), G. Kreisel reviewed the book which I shall examine in this essay. He closed this review by commenting: "now it seems to me to be a surprisingly insignificant product of a sparkling mind". Although one can sympathize with this comment, it is not entirely accurate. It is not surprising that the book as a whole seems insignificant because it is a posthumous selection of remarks from his note-books. These remarks had been written over a period of at least seven years, 1937 to 1944. In the text they are organized according to their date of composition, rather than according to a connecting theme, though I must confess that in my essay I treat his work as a treatise. Furthermore, many of the remarks are very obscure. They are like the expressions of an insight which one jots down in a note-book and then needs several pages to *unpack* it. Unfortunately, Wittgenstein did not *unpack* his insights. I submit that my theory of strict finitism which is presented in Chapters II and III is an *unpacking* of several of his remarks, and of those remarks which express his deepest insight on mathematics. There is no doubt that if one thought that the only significant remarks on the foundations of mathematics occurred in mathematical journals, such as the *Journal of Symbolic Logic,* then Wittgenstein's *Remarks* would seem surprisingly insignificant. There is no discussion of the technical problems of axiom systems except to say that they are philosophically unimportant. Only an elementary knowledge of symbolic logic is required to read his attacks on formal proofs in mathematics and the attempt to show that mathematics is logic in *Principia Mathematica*. Although Wittgenstein's *Remarks on the Foundations of Mathematics* is insignificant as a mathematical or metamathematical treatise, it is not fair to expect it to be one. A brief glance through it immediately shows that it is not at all like the technical articles on that branch of mathematics called "the foundations of mathematics". I hope that my essay is no more technical than Wittgenstein's *Remarks*.

However, many of Wittgenstein's remarks have surprising philosophical significance. In my opinion the most significant points are the following three. Wittgenstein accounts for the necessary truth of simple geometry and the arithmetic of small numbers without any trace of platonism; this is what I call his strict finitism. He finds the proper place for self-evidence; in other words, he recognizes the importance of not being able to conceive of the denial of a proposition without using inconceivability of the denial as a criterion for or definition of necessary truth. Thirdly, he shows, by non-mathematical means, the mathematical and philosophic irrelevance of attempts such as *Principia Mathematica,* to reduce mathematics to logic. I do believe that these *Remarks* contain a conclusive refutation of Logicism.

This essay discusses the same topic as my dissertation: *An Examination of Ludwig Wittgenstein's Remarks on the Foundations of Mathematics,* viz., Wittgenstein's *Remarks*. The only similarity between the two, however, lies in the presentation of strict finitism. In my dissertation, I tried to uncover an implicit metaphysics in Wittgenstein's strict finitistic philosophy of mathematics and then tried to refute Wittgenstein by refuting this metaphysics which I attributed to him. In this essay my aim is not to criticize Wittgenstein. My aim has been primarily to develop some lines of thought suggested in Wittgenstein's *Remarks* in hope that such developments may make reading these remarks of Wittgenstein's easier and more profitable.

Throughout my essay I abbreviate Wittgenstein's *Remarks on the Foundation of Mathematics* as *RFM*. *RFM* is broken up into five books: I, II, III, IV, and V. Each book is divided into remarks numbered by arabic numerals. Thus I shall refer to specific passages, such as remark 22 in Book II by citing book number and remark number, eg., II-22. I abbreviate Wittgenstein's *Philosophical Investigations* as *PI*. Most of the papers on *RFM* which I consider are reprinted in: *Philosophy of Mathematics,* Selected Readings, edited by Paul Benacerraf and Hilary Putnam (Prentice Hall, 1964). I abbreviate this book of papers as *BP*. I hope that the absence of footnotes facilitates reading.

TABLE OF CONTENTS

I

PHILOSOPHICAL PROBLEMS AT THE FOUNDATIONS OF MATHEMATICS

> I have not yet made the role of miscalculating clear. The role of the proposition: "I must have miscalculated". It is really the key to an understanding of the 'foundations' of mathematics. (II-90)

1. It is not obvious that the role of an admission of miscalculating is the key to an understanding of the foundations of mathematics. What is the role of a proposition? What is meant by "foundations of mathematics"? How could an understanding of such an apparently minute point, namely an admission of miscalculation, be so helpful in such an apparently large topic, namely the foundations of mathematics? Yet, Wittgenstein is basically correct. To show that he is correct, I shall first describe what I believe Wittgenstein meant in this passage by "foundations of mathematics". Then, I shall explain how understanding something which you could call the role of "I must have miscalculated" provides answers to some of the crucial questions about the foundations of mathematics. However, my main goal in this chapter is not to explain this passage about miscalculating. Rather, I hope, while explaining this passage about miscalculating, to reach the main goal of this chapter – to explain what I consider to be an investigation into the foundations of mathematics.

In this essay, by "foundations of mathematics" I mean, and I believe that in *RFM* Wittgenstein meant, philosophic foundations of mathematics as opposed to mathematical foundations of mathematics. In IV-52 Wittgenstein wrote:

The philosopher must twist and turn about so as to pass by the mathematical problems, and not run up against one, – which would have to be solved before he could go further.

His labour in philosophy is as it were an idleness in mathematics.

In V-13 he wrote:

The *mathematical* problems of what is called foundations are no more the foundations of mathematics for us than the painted rock is the support of a painted tower.

But what is the philosophy of mathematics or an investigation into the philosophic foundations of mathematics? And what, if anything, are mathematical foundations of mathematics? In my opinion, someone is investigating the philosophic foundations of mathematics when he tries to answer one or more of the following questions. (I have obtained the idea of the philosophy of mathematics being based on questions like these from the Introduction of Stephen Körner's *The Philosophy of Mathematics* (London, 1960).

Group I: Questions on the ontology of non-applied mathematics.

(i) When we do non-applied mathematics are we talking about some objects?

(ii) When we do non-applied mathematics are we talking about non-ordinary objects?

(iii) If we are talking about non-ordinary objects when we do non-applied mathematics, what are they like and how do we get knowledge of them?

Group II: Questions on the epistemology of non-applied mathematics.

(i) Are there truths in non-applied mathematics?

(ii) If there are truths in non-applied mathematics, are they necessary truths?

(iii) If the truths of non-applied mathematics are necessary truths, what does it mean to say that they are necessarily true in addition to their being merely true and how do we come to know that they must be true?

Group III: Questions designed to check whether or not answers to the preceding questions on the ontology and epistemology of non-applied mathematics enable the philosopher to account for applied mathematics.

(i) Does a bit of non-applied mathematics become applied mathematics by making it be about objects if it were about no objects as non-applied mathematics? Or, if it were about objects when non-applied, does it become applied by making it be about a different kind of object? If non-applied mathematics is about the same objects as applied mathematics, what is the difference between applied and non-applied mathematics?

(ii) Does a proposition of non-applied mathematics become applied by making it be a truth about something if it were not true when non-applied? Are propositions of applied mathematics necessarily true? If propositions of applied mathematics and non-applied mathematics are both necessarily true, what differentiates applied from non-applied mathematics?

2. Before I state what I mean by an investigation into the mathematical foundations of mathematics, I shall comment extensively on these questions leading to a philosophy of mathematics. What is meant by "applied" and "non-applied"? I cannot now say precisely what is the difference between non-applied and applied mathematics. To state precisely the difference between non-applied and applied mathematics is, as is clear from Group III, one of the problems of the philosophy of mathematics. However, the following distinction will suffice for our present purpose of illustrating these groups of questions. A person has done applied mathematics if he makes or accepts a non-mathematical statement and he has obtained it by doing mathematics. The mathematics that he used to conclude his non-mathematical proposition is applied mathematics.

For instance, a person may say "This box of soap weighs three pounds." The statement that a box of soap weighs three pounds is not a mathematical statement. Assume that we ask this person how he arrived at this conclusion because we know that he did not weigh the box. If he answers that he figured it out because he knew that this box and a smaller one, which weighs one third what it weighs, together weigh four pounds, he has done applied mathematics.

Of course, when doing his applied mathematics, the person gives mathematical expression to facts about his environment. In our example, the man would have expressed the relevant facts in the system of equations: $b_1 + b_2 = 4$ and $b_2 = b_1/3$. In this system b_1 stands for the weight of big box and b_2 for the weight of the smaller box. Similarly, conclusions about the pressures of gases, accelerations of objects towards one another, and the amount of interest earned in four months are drawn by having a mathematical expression for some facts of nature and then computing the non-mathematical conclusion from them. (I include statements of initial conditions such as $T = 17°C$ as mathematical expressions of natural facts.) Such conclusions about pressures, etc., we would normally say are obtained as a result of doing applied mathematics. According to my characterization of applied mathematics, the drawing of such conclusions would be typical applications of mathematics. So I take my characterization of applied mathematics as justified. Note, however, that I do not consider giving mathematical expression to natural facts applied mathematics, although it may be called applying concepts used in mathematics. In this I also agree with common sense because to state laws of nature mathematically is not considered to be a function of mathematics; it is considered to be a function of natural science.

An advantage of characterizing applied mathematics as the mathematics that a person uses to arrive at a non-mathematical conclusion is that I do not beg the philosophic question of whether the subject matter and modality of mathematical propositions are the same or different in applied and non-applied mathematics. (By modality of a proposition I mean whether it is necessarily true (or false) or contingently true (or false). Very likely the person in our example concluded that the big box weighed three pounds by solving the system of equations: $b_1 + b_2 = 4$ and $b_2 = b_1/3$. His solving of this system was doing applied mathematics. However, in solving this system we do not know whether or not he has done the same as he would have done if he were solving this system as a problem in non-applied mathematics. For instance, when he solved this system he may have been doing the same as he did as a student in an algebra class or he may have done something quite different.

3. A person has done non-applied mathematics if he has done mathematics but has not done it for the purpose of drawing a non-mathematical conclusion. By doing mathematics I have in mind proving a theorem of mathematics or computing the value of a function. Since computing the value of a function, eg., computing that the cube root of 27 is 3, can be regarded as proving a low level theorem, eg., $\sqrt[3]{27} = 3$; I would say that doing non-applied mathematics is proving mathematical theorems. What, then, is mathematics? Mathematics is what is considered to be mathematics by university departments of mathematics and editors of mathematical journals.

4. My characterization of mathematics requires comment on two points. The first is that I admit that what we ordinarily call computing or calculating can be non-applied (pure) mathematics as much as proving the fundamental theorem of arithmetic is doing pure mathematics. (Establishing that $15 \times 13 = 195$ is a typical example of what we ordinarily call computing or calculating.) In this I agree with Wittgenstein. The opening quotation of this chapter hints that he regarded calculating as an important part of non-applied mathematics. Also, a reading of *RFM* will show that he is trying to gain a philosophic understanding of mathematics, to answer the three groups of questions listed above, primarily by asking these questions about that level of mathematics we call calculating. Indeed in his "Mathematics and the 'Language Game'" (*BP* 481-490), Alan R. Anderson went so far as to criticize Wittgenstein for identifying mathematics with calculating procedures. But Wittgenstein did not make the factual error of identifying mathematics with calculating procedures. For instance, in IV-46 he admitted: "'Mathematics' is *not* a sharply delimited concept." Thus he would not believe that he could identify mathematics, which he cannot clearly identify, with calculating. Also in II-46 he explicitly claimed: "mathematics is a MOTLEY of techniques of proof".

5. If Wittgenstein made any kind of error of identifying mathematics with calculating, it was the error of holding that a philosophic understanding of calculating both applied and non-applied, gives a philosophic understanding of mathematics on any level. I do believe that Wittgenstein thought that a philosophic understanding of calculating would provide a philosophic understanding of all of mathematics. To support this, I again refer to the opening quotation and his practice in *RFM*. I would also call attention to IV-16 in which he wrote:

> Strangely, it can be said that there is so to speak a solid core to all these glistening concept-formations. And I should like to say that that is what makes them into mathematical productions.

In our development of his philosophy, we shall see that Wittgenstein regards mathematics as a way of forming concepts. However, an important point to make here is that the assertion of there being this solid core should not be read as contradicting the claim that there is a motley of mathematical techniques of proof. It should, I submit, be taken as a suggestion that there is one facet of all mathematical thinking such that, if properly understood in any one of the manifold branches of mathematics, this understanding would provide a philosophic understanding of all other branches of mathematics. Now it is a philosophic error to hold that a philosophic understanding of calculating suffices for a philosophic understanding of mathematics only if there is no such solid core. So whether or not Wittgenstein made a philosophic error by focusing his attention primarily on calculating can be judged only after we have his philosophic examination of calculating.

6. The second point of comment on my characterization of non-applied mathematics is the loose characterization of mathematics as that which mathematicians call mathematics. Such a characterization is certainly compatible with Wittgenstein's admission that "mathematics" is a vague, - not sharply delimited –, concept. I interpret "vague" or "not sharply delimited "as meaning that there are no necessary conditions for applying "mathematics".

Also in V-26 Wittgenstein wrote:

> Mathematics is, then, a family; but that is not to say that we shall not mind what is incorporated into it.

This use of "family" indicates that Wittgenstein held that "mathematics" is a family concept in the way in which he showed that the concept "game" is a family concept in 65-77 of his *Philosophical Investigations*. I would say that a concept, C, is a family concept if we cannot state any one feature necessary for everything that we call a C to be a C but yet there are limits on what can be called a C.

7. But is not an admission that "mathematics" is a family concept incompatible with the claim that there is a solid core to mathematics? There is no incompatibility because what people now call mathematics may be so called without there being any necessary condition that they use for calling it all mathematics. Nonetheless there may be one common feature of everything which is now called mathematics such that, if properly understood, it would provide a philosophic understanding of everything which is NOW called mathematics. However, there is no claim that this common feature is a necessary condition for something to be mathematics now. Nor is there any claim that this common feature is something we can start to use as a necessary condition for being mathematics. We cannot say how "mathematics" will be used in the future. Five hundred years from now people may label activities as "mathematics" which do not have the common feature we found. In such a case, we would need a new philosophy of mathematics; but not a new definition of "mathematics".

The importance for this essay of using my loose characterization of mathematics is that by using it we shall follow Wittgenstein in our attempts to gain philosophic understanding of mathematics. We shall try to gain philosophic understanding of what is now called mathematics without trying to foreclose on what can be called mathematics. Or, as Wittgenstein put it in IV-52:

> It is not that a new building has to be erected, or that a new bridge has to be built, but that the geography, *as it now is*, has to be judged.

And a bit later in the same passage he wrote:

> Even 500 years ago a philosophy of mathematics was possible, a philosophy of what mathematics was then.

(For a similar statement see II-81.) So in Wittgenstein's *RFM* we should expect no definitions of "mathematics". Similarly, we should expect no altering of mathematics as it is actually done, eg., re-doing mathematics in symbolic logic. He wanted to understand mathematics as it is actually done now but he does not want to say that what is done now is or is not REALLY mathematics. He wants only to understand whatever it is that mathematicians are now doing. If they do something different in the future, it will be up to philosophers of that age to understand what it is that their mathematicians are doing.

8. I agree with Wittgenstein that what we want to understand is mathematics as it now is. Thus I agree with Wittgenstein that in the philosophy of mathematics we should leave it as it is. If we change it, we will be studying something else. Wittgenstein stated this position very well in PI-124 where he wrote:

> Philosophy may in no way interfere with the actual use of language; it can in the end only describe it.
>
> For it cannot give it any foundation either.
>
> It leaves everything as it is.
>
> It also leaves mathematics as it is, and no mathematical discovery can advance it. A "leading problem of mathematical logic" is for us a problem of mathematics like any other.

Of course, when he claimed that no mathematical discovery can advance the philosophy of mathematics, he was claiming that a new discovery such as the discovery that the predicate calculus is recursively undecidable will not help us understand mathematics as it is now done since this new discovery is only part of the current mathematics to be understood. However, he did not mean that mathematical advances are totally irrelevant for philosophy because the advances may give new mathematics to philosophize about. Thus Michael Dummett is a bit inaccurate when he criticizes Wittgenstein for sharply segregating philosophy from mathematics on p. 492 of *BP* in his article "Wittgenstein's Philosophy of Mathematics" which is reprinted on pp. 491-509 of *BP*.

9. But now that I have said so much about gaining a philosophic understanding of what mathematicians are doing, I should return to complete my discussion of what it is to try to gain a philosophic understanding of mathematics. This requires that I return to commenting on my three groups of questions. What is question (Ii) supposed to bring out? It is supposed to bring out whether or not a philosopher believes that an essential step in coming to a philosophic understanding of a mathematical proposition is to describe the truth conditions for it. In his "Wittgenstein's Philosophy of Mathematics", Michael Dummett suggests that the fundamental division in the philosophy of mathematics stems from whether or not a philosopher holds that a description of truth conditions rather than a description of assertibility conditions is the proper way to explain the meaning of a mathematical proposition. Although I disagree with Dummett's view that Wittgenstein holds a conventionalistic interpretation of necessity, Dummett is correct in holding that these two ways of explaining the meaning of mathematical propositions lead to the fundamental philosophic division between those who hold that mathematical truths tell us the order and connexion of some objects and those who either hold that there are no truths in mathematics or hold that mathematical truths do not tell us of the order and connexion of some objects. For this reason, I think that an investigation into the philosophical foundations of mathematics should begin with question (Ii). Of course, a person's initial answer to (Ii) may need to be revised if it gives him difficulty on the other questions or has repugnant philosophic consequences later on.

But what does it mean to explain the meaning of a sentence? The answer is in Russell's Introduction to Wittgenstein's *Tractatus Logico-Philosophicus*. Russell claims that Wittgenstein in the *Tractatus* is concerned with the question: "what relation must one fact (such as a sentence) have to another in order to be *capable* of being a symbol for that other?" I believe that this question of the *Tractatus* is a special case of a more general question which we can call, misleadingly as we shall see, explaining the meaning of a sentence. The more general question is: How does a sentence do what people use it to do? Why worry about how a sentence can be used to do what it is used to do? A philosopher has such a worry if he characterizes some use of a sentence in a particular way, has a preconceived idea on how this use as he characterized it is carried out, but yet the conditions he believes must be met for it to be so used do not obtain. A fine example of such a problem was Russell's worry about "The present King of France is bald." He could imagine characterizing a use of this sentence as fact stating. According to his preconceived idea of what must obtain for this sentence to be used to state a fact the term "The present King of France" should denote some individual. But it does not. Thus Russell had the problem of how "The present King of France is bald" could be used as a fact stating sentence. Ethics provides another example of this kind of problem. Someone might characterize uses of sentences such as "Physical health is good" as fact stating but also believe that "good" should stand for some observable property if the sentence is used to state a fact. Consequently, such a philosopher would ask how ethical sentences can do the job, viz., state facts, which he believes that they do. Another philosopher may characterize all uses of sentences as expressions of thoughts but, because of preconceived ideas of what thoughts are like, be unable to understand how audible or visible sentences can carry thoughts. I believe that with ingenuity one could restate most, if not all, philosophical problems as wonder about how a kind of sentence can do the job it is believed to do, given a philosopher's preconceptions on how that job can be done.

Our main concern here, however, is with problems about mathematical objects and necessity, in other words, problems about the ontology and epistemology of mathematics. Such problems can be formulated as doubts about how sentences can do jobs which they have been said to do. For instance, mathematical sentences may be characterized as primarily devices for stating facts. If one has the preconceived idea that a sentence can be used to state a fact only if some of the terms in it are used to refer to observable individuals and properties which are arranged in a way in which they do not have to be arranged, he will wonder how mathematical sentences can state facts, since if they are about anything they are about how unobservable individuals and properties must be arranged. Actually, I have designed my three groups of questions as attempts to determine in what way, if any, mathematical sentences can be used to state facts, granted that it is extremely doubtful that they say how observable objects happen to be arranged. (Of course, once one has the problem of how a sentence can do the job it allegedly does, he should also be prepared to ask whether the sentence really is used to do the job it is said to be used for or to modify the preconceived ideas on what must hold for this job to be done.) They are two ways for a philosopher to handle problems due to his characterization of the role a sentence plays and his preconceived idea of what must obtain for that role to be played. The first way corresponds to explaining meaning in terms of truth conditions while the second corresponds to explaining meaning by describing assertibility conditions.

10. Explaining the meaning of a sentence in terms of its truth conditions is to explain how the sentence can be used to state a fact when it is doubtful that it can state a fact, given preconceived ideas of what it is to state a fact. For instance, it is doubtful that the kind of conditions which entitle us to say "Peter is blond" exist when we assert "7 is prime". So, a philosopher who holds that "7 is prime" states a fact will wonder how it can do so. But, let us say that he is not willing to give up the belief that "7 is prime" states a fact. Nor, is he willing to change radically his conception of what there must be for a sentence to state a fact. Let us say that our philosopher holds that for a sentence to state a fact the non-logical signs must stand for observable individuals or properties and the logical signs must be interpreted so that the sentence takes on truth values depending upon how the objects referred to are arranged. Thus, he holds that "Peter is blond" states a fact only if "Peter" is used to stand for a human male, "blond" is used to stand for the property of having yellowish hair, and the logical word "is" is interpreted as saying that the man Peter and the property blond happen to be arranged so that the hair of Peter exemplifies the property of being blond. Now he

cannot, let us assume, say that "7" stands for an observable individual and that "prime" stands for an observable property, and he cannot say that it merely happens to be the case that 7 is prime. (He cannot say these things because he wants to agree with ordinary usage.) Still he wants to say that "7 is prime" states a fact only if "7" refers to some individual, "prime" to some property, and "is" says that some kind of exemplification occurs. So this philosopher has a phenomenon to explain. The phenomenon he has to explain is that "7 is prime" is capable of stating a fact. Very likely he would account for this alleged fact by developing a theory which has non-observable objects necessarily connected, i.e., a species of platonism. The important point for this essay, however, is that he has explained meaning, – explained how a sentence can be used to state a fact –, by developing a theory about how some objects are connected in order to provide truth conditions for fact stating uses of the sentence.

11. There is a second way in which a philosopher can handle this kind of problem, which arises when his characterization of the role a sentence plays and his preconceived ideas on how that role is played, clash with the way in which the sentence is actually used. He can take the position that as a philosopher he cannot refer to something beyond language to account for how language operates. He would hold that it is the task of empirical science to discover what and how things beyond language enable language to function as it does. All that a philosopher can do to remove his doubts about how language functions is to become clear on how it does function. To accomplish this he shall have to drop his original characterizations of use and preconceived ideas on how language can be used as he had characterized it. He must simply be clear about what goes on when people make assertions. (At this point, it is extremely important to emphasize that describing what goes on when people make assertions is to do far more than merely describe what is said and to point out what forms of words people would regard as bizarre. In *RFM*, we shall find that describing what goes on when people make mathematical assertions consists primarily of calling attention to non-linguistic facts, viz., physical, psychological, and sociological facts.) After

he is clear about what goes on when people make assertions, questions on why and how this happens will be questions for empirical science. So in effect his philosophic work presents data for scientific explanation. However, the purpose of the philosophic description of what goes on when people use language is not to present data for science. Its purpose is to answer the philosophic question on how a sentence can do what it is used to do. But the philosophic question is answered by removing or trying to remove the question rather than by developing a theory to answer it.

As a result of becoming clear about how language is used we, it is hoped, will no longer ask for non-scientific explanations of how it works. Even if we continue to ask, as we probably shall, how language can do what it does, we shall not answer such questions because they lead us into non-empirical science. Instead, we shall continue to try to understand better what is being done in the particular use in question. We shall try to make the question disappear; we shall not try to build on the question in the sense that we shall not try to build a theory about some objects in order to answer it. (That is, we shall do these things if we adopt the second way of handling these problems.) Similarly, we do not try to build on a disease by developing ways of life to foster it; we try to remove it. As Wittgenstein said on occasion, for instance in *PI*-255: "The philosopher's treatment of a question is like the treatment of an illness." This means that the philosopher is supposed to try to make his question disappear without traces, i.e., explicit answers. Of course, the presupposition of this way of handling philosophical problems is as Wittgenstein put it so well in *PI*-126: "Since everything", what is said and the conditions for saying it, "lies open to view there is nothing to explain. For what is hidden, for example, is of no interest to us."

12. How did Wittgenstein explain the meaning of sentences? In *RFM* he used the second way. But it was not always so with him. Russell notes in his Introduction to the *Tractatus*, and as Wittgenstein himself admitted in *PI*-114, Wittgenstein of the *Tractatus* characterized the role of sentences as fact stating. (Thus we get the restricted question quoted from Russell in section 9.) As he continued his philosophical investigations, Wittgenstein re-

jected the assumption that the essential role of a sentence is to state facts. This rejection is brought out in *PI* 18-24. But I think that a more significant alteration of his philosophical views occurred when he came to the conclusion that the "how" in "How does a sentence do what people use it to do?" asks for a description rather than for an explanation in terms of some theory. (It makes no difference if the philosopher calls his theory meaningless as Wittgenstein did at the end of his *Tractatus*; he still has the theory.) On page 18 of his *Blue Book* Wittgenstein wrote:

> Philosophers constantly see the method of science before their eyes, and are irresistably tempted to ask and answer questions in the way science does. This tendency is the real source of metaphysics and leads the philosopher into complete darkness. I want to say here that it can never be our job to reduce anything to anything, or to explain anything. Philosophy really *is* 'purely descriptive'.

This alteration is more significant than his rejection of the mistake that sentences are essentially fact stating devices because as a result of these observations he completely changed his way of doing philosophy. He changed from developing theories about some objects, which make language work as it does, to describing how language actually works. The point of this brief sketch of Wittgenstein's philosophical development is to account for his philosophizing by describing language and conditions for its use. He changed in order to avoid the kind of theory building in his *Tractatus,* since such theorizing only produces more puzzles when we try to make the theory clear.

Unfortunately, the method of describing assertibility conditions has a philosophic point only if one is trying to remove some preconceived ideas on how language is used and some theories on how it can be used this way. In other words, the kind of philosophy Wittgenstein did later in his career makes sense only as an antidote for his earlier philosophy or other metaphysical schemes. If usage is not described for the sake of removing the questions which give rise to metaphysical theory building, the description becomes the mere, rather casual, compilation of data for scientific explanation. So, unfortunately *RFM* will be of value only to someone who has developed or is tempted to develop a metaphysical scheme to explain how we can have the mathematics that we actually have. It will not help an unprejudiced mind.

13. Let us return to ask again exactly what question (Ii) is supposed to bring out. It is supposed to bring out whether or not the philosopher has enough of a theory of language, – non-empirical most likely –, to try to explain the meaning of mathematical sentences in terms of truth conditions. It should be clear now, though, that to explain meaning does not require one to talk about something called meaning; one is only trying to understand how a sentence can be used to do what it is believed to do.

If one answer (Ii) negatively, he can skip to (IIi). His negative answer has determined that he will not answer questions from Group II by means of discussing truth conditions. However, if he answers (Ii) affirmatively, he has not yet forced himself to hold that there are some special objects constituting truth conditions for mathematical propositions. But let us follow up the results of affirmatively answering (Ii), viz., admitting that mathematical statements tell us how it is with some objects. It is well to follow up the results of holding that mathematical sentences have truth conditions because only when we see what else we may have to hold when we hold that mathematical sentences have truth conditions can we appreciate why anyone would claim that they do not. Remember, though, that when I talk about truth conditions I mean some objects, i.e., individuals and properties, referred to in the sentence such that if they are arranged in a certain way the statement made with the sentence is true; otherwise not. I do not mean by "truth conditions" simply any conditions which entitle a person to apply "true" to a statement or proposition. For in this very general sense of "truth conditions" any sentence which can be used to make a statement or to express a proposition, which people call true or false, has truth conditions. If one admits that mathematical sentences have truth conditions, he must also answer (Iii). Are the objects that we are talking about in non-applied mathematics ordinary objects? Admittedly, there is no sharp distinction between ordinary and non-ordinary objects. In most localities, a man with black teeth is out of the ordinary.

But for purposes of this essay, what do I mean by "non-ordinary objects"? If someone claims that there are some individuals and relations whose order and connection can be discovered only by doing mathematics, he has claimed that there are, for purposes of this essay, non-ordinary objects. I have in mind here someone claiming that the objects which they talk about in mathematics cannot be dealt with by common sense observation because they lack the secondary qualities of color, odor, visual size, shape, and feel while adding that they cannot be dealt with by the method of physical science because they do not have properties definable in terms of length, mass, and time. Of course, a person who claims that there are mathematical objects, as above, would also claim that they cannot be studied by the social sciences since these alleged mathematical objects are non-temporal individuals and relations. I do not think that it is unfair to label atemporal, unobservable objects "non-ordinary". Nor, do I think that it is inaccurate to label as platonistic a view of mathematics which holds that there are truth conditions for mathematical sentences consisting of such non-ordinary objects, as the term "platonistic" is presently used in a non-technical slightly abusive way.

14. I shall here add a note about the use of the term "platonist" in this essay. In this essay, I shall follow Dummett's explication of the term in his "Wittgenstein's Philosophy of Mathematics". Dummett wrote, on p. 492 of *BP*, "According to Platonism, mathematical objects are there and stand in certain relations to one another, independently of us, and what we do is to discover these objects and their relations to one another." Dummett's statement of platonism is not too different from Bernays' statement of absolute platonism in his "Sur le Platonisme dans les Mathématiques" an English translation of which appears in *BP* on pp. 274-286. According to Bernays', on p. 277 of *BP*, absolute platonists postulate "the existence of a world of ideal objects containing all the objects and relations of mathematics". Earlier, on p. 275 of *BP*, Bernays noted that they view "the objects as cut off from all links with the reflecting subject". Except, of course, they can be known by the reflecting subject; they simply can exist without being known and they are not affected by being known. On p. 277 of *BP*, Bernays contrasts absolute

platonism with what he calls a more restricted platonism. According to Bernays a restricted platonist talks about there being a domain of individuals and properties which are studied exclusively by mathematics but he does not claim this domain "to be more than, so to speak, an ideal projection of thought". In effect, a restricted platonist does not hold that mathematical objects exist independently of human thinking. For instance, an absolute platonist would hold that the totality of integers exists independently of anyone thinking about them. A restricted platonist would hold that the totality of integers exists only if people think of such a totality. However, after he has posited, – reified –, the totality of integers, a restricted platonist would talk about particular integers existing independently of being thought about. More generally, a restricted platonist admits that domains of mathematical objects are mind-dependent but that within these mind-dependent domains he can talk about there being objects unknown and unthought of.

My purpose in talking about platonism is twofold. Wittgenstein is charged with dogmatically rejecting platonism. To understand this charge we must understand what Wittgenstein would have meant by "platonism" had he used the term. Second, Bernays maintains that the occurrence of the antinomies, e.g., Russell's paradox involving the class of all classes which do not belong to themselves, show that absolute platonism is untenable. If Bernays were correct on this second point, Wittgenstein was wrong about how to philosophize about mathematics. Recall, from section 8, that Wittgenstein held that mathematical results are philosophically irrelevant except insofar as they provide more material to philosophize about.

15. Wittgenstein would have found both absolute and restricted platonism philosophically repugnant. In IV-5 he wrote:

Someone makes an addition to mathematics, gives new definitions and discovers new theorems – and in a *certain* respect he can be said not to know what he is doing. – He has a vague imagination of having *discovered* something like a space (at which point he thinks of a room), of having opened up a kingdom, and when asked about it he would talk a great deal of nonsense

Later in IV-5, he wrote:

> In other words: if someone believes in mathematical *objects* and their queer properties – can't he nevertheless do mathematics? Or – isn't he also doing mathematics.

I think that these remarks make it clear that Wittgenstein rejects belief in mathematical objects and that he was not inclined to accept a mathematicians account of what he does in mathematics. His disinclination to accept mathematicians' philosophy of what they are doing is a clue that he rejected restricted as well as absolute platonism. (I take restricted platonism as just the kind of position that a mathematician would advance if he thought that he was talking about objects when he did mathematics, but was a bit timid about advancing his position.) Restricted platonism would have been repugnant to Wittgenstein because talk about mathematical domains being created by the mind reveals the belief that one needs a theory about some objects to account for their use of sentences in mathematics. Also, mind-dependent domains with mind-independent members are almost as mysterious as the domains of the absolute platonists, and I think the above remarks make it clear that Wittgenstein was loathe to talk about mysterious entities, be they in or out of the mind. The final paragraph of IV-5 makes it more certain that he was opposed to talking about objects for mathematical propositions, even if it is conceded that they exist only in the mind. In this last paragraph, he suggested that the phrase "ideal object" is misleading. Finally, if we turn to *PI*-254 where he wrote:

> what a mathematician is inclined to say about the objectivity and reality of mathematical facts, is not a philosophy of mathematics, but something for philosophical treatment . . .

We have even more support for my contention that he would consider a statement of restricted platonism as problematic as a statement of absolute platonism. So, although there is a difference between absolute and restricted platonism, in the rest of this essay I shall talk about platonism as absolute platonism. I shall do this because Wittgenstein does not precisely distinguish between the two. He probably would have regarded restricted platonism as simply a hopeless attempt to patch up a bad position.

16. Why does Bernays claim that the mathematical results of the antinomies show that the philosophical position of absolute platonism is untenable? He is rather terse on this point. On p. 277 of *BP*, Bernays writes:

> The essential importance of these antinomies is to bring out the impossibility of combining the following two things: the idea of the totality of all mathemathical objects and the general concepts of set and function; for the totality itself would form a domain of elements of sets, and arguments and values for functions.

I think that Bernays' argument is as follows. If someone regards the totality of mathematical objects as being the same kind of thing as mathematical objects, then he accepts that the totality of mathematical objects can be dealt with, as other mathematical objects are dealt with. This means that the totality of mathematical objects will be dealt with as if it can belong to a set, be an argument of a function, or be a value of a function. If someone accepts that the domain of all mathematical objects can be dealt with as other mathematical objects, he has the antinomies. If someone is an absolute platonist he holds that the domain of all mathematical objects is the same kind of thing as particular mathematical objects. Thus someone who is an absolute platonist has the antinomies. But since no one can accept the antinomies no one can, rationally, be an absolute platonist. For instance, an absolute platonist may say that the domain of classes is the domain of mathematical objects, where class is understood intuitively as all and only those objects satisfying a certain condition. If they regard the domain of classes as the same kind of thing as other classes, they may easily slip into the view that the domain of classes is the class of all classes and that this class can be dealt with as other classes are dealt with. For instance, they may split the class of all classes into two parts, viz., those that belong to themselves and those that do not. Once they make this split, they have the Russell paradox as soon as they realize that their belief that the class of all classes can be split into two parts requires them to determine whether or not the class of all classes which do not belong to themselves does or does not belong to itself.

Now I believe that Bernays is correct on all his premisses except for the one saying that if someone holds that the domain of all mathematical objects is the same kind of thing as mathematical objects he also holds that the domain of all mathematical objects can be treated as another mathematical object. This premisse is unclear on exactly what it is to regard two things as being of the same kind. In this context in which we are discussing absolute platonism, the fair interpretation of "A is of the same kind as B" is "A and B both exist independently of (dependent upon) being thought about or being known." With this interpretation, there is no reason for charging that an absolute platonist holds that the domain of mathematical objects is itself a mathematical object like the other mathematical objects. All that an absolute platonist need hold is that both the domain of mathematical objects and mathematical objects themselves exist outside and independently of minds. Indeed, an absolute platonist can regard the antinomies as a mathematical proof of the fact that the mind-independent domain of mathematical objects is mathematically different from the mathematical objects with which it is ontologically similar. So, if I am right, the mathematical discovery of the antinomies did not resolve the philosophic issue between absolute and restricted platonism. Thus we do not here have a counter-example to Wittgenstein's claim that mathematical results are philosophically irrelevant.

17. I believe, and I think that Wittgenstein believed, that if we do accept that mathematical sentences have truth conditions, i.e., give an affirmative answer to (Ii), we will be forced to platonism, i.e., giving an affirmative answer to (Iii). I believe this because as soon as we admit that there are mathematical objects, integers for example, we immediately encounter all the hackneyed arguments that they cannot be ordinary objects. For instance, one may face an argument that the number two could not be an observable individual because one can weigh an observable individual but it makes no sense to talk about weighing the number two. I

call these arguments hackneyed because it takes no real understanding of any subject to make these obvious observations about what can be said. No insight is provided by these observations, and it is surprising that anyone could think that an important fact about the world could be established by such casual observations. But note that I am not saying that such ways of arguing are trivial if we are NOT doing metaphysics. If we are merely trying to be clear about what is said, such facts on what can be said may be very relevant. This way of arguing would be what Wittgenstein in PI-127 calls assembling reminders for a particular purpose. Such arguments are relevant if they remind us how number words are actually used; but they do not show anything about the nature of truth conditions for mathematical sentences.

18. It is more difficult to establish that Wittgenstein denied that mathematical sentences have truth conditions in order to avoid platonism. But it is easy to establish that Wittgenstein denied that mathematical sentences have truth conditions. I mean that it is easy to establish that Wittgenstein rejects a view that there are mathematical facts to provide truth conditions for statements made in mathematics. In V-15 he suggested:

it takes mathematics to define the CHARACTER of what we are calling a 'fact'!

He went on to say in V-15:

> It is interesting to know *how* many vibrations this note has! But it took arithmetic to teach you this question. It taught you to see this kind of fact.

Later in V-15 he challenged himself with:

> To be practical, mathematics must tell us facts.

Since he had no doubt about the practicality of mathematics, he must face the challenge. His reply was:

> But do these facts have to be the *mathematical* facts? – But why should not mathematics, instead of 'teaching us facts', create the form of what we call facts?

These remarks support, if they do not establish, that Wittgenstein did not accept truth conditions, – mathematical facts –, for mathematical sentences.

Indeed Bernays, on p. 513 of *BP*, in his article "Comments on Ludwig Wittgenstein's *Remarks on the Foundations of Mathematics*", which is reprinted on pp. 510-528 of *BP*, charges that Wittgenstein uncritically adopted "a preconceived philosophical view . . ., that there can exist but one kind of factuality: that of concrete reality". Bernays also writes in this article, on p. 513 of *BP*,

> The fundamental division between the sphere of mathematics and the sphere of facts appears in several passages in the book. In this connection Wittgenstein often speaks with a certitude which strangely contrast with his readiness to doubt so much of what is generally accepted.

Bernays is, I believe, certainly correct that Wittgenstein, with uncharacteristic directness, proclaims that there are no mathematical facts. Bernays cites such striking remarks as:

"If you know a mathematical proposition, that's not to say you yet know anything" and "the mathematical proposition is only supposed to supply a framework for a description" from V-2. (One can find more support for the claim that Wittgenstein denied that there were mathematical facts in I-155.)

19. I do not, however, accept Bernays' complaint that Wittgenstein uncritically accepted the generally accepted (philosophical) view when he made this sharp distinction between factual and mathematical propositions. To the contrary, Wittgenstein was well aware of the fact that many people, both philosophers and nonphilosophers, are inclined to talk about mathematical objects and, hence, mathematical facts. He realized that our ways of speaking in mathematics suggests that there are mathematical facts. In III-11 he wrote:

> Arithmetic as the natural history (mineralogy) of numbers. But *who* talks like this about it? Our whole thinking is penetrated with this idea.

(Bernays also notes this passage.) In Appendix I-4, Wittgenstein wrote:

> We are used to saying "2 times 2 is 4", and the verb "is" makes this into a proposition, and apparently establishes a close kinship with everything that we call a 'proposition'. Whereas it is a matter only of a very superficial relationship.

So Wittgenstein did not unconsciously assume the popular position on mathematical facts, although I do grant that he did not directly argue for the position that mathematical sentences do not have truth conditions.

However, although he did not directly argue that mathematical statements are not statements of fact, he did show that mathematical statements do not state experimental facts. And his discussion of whether or not mathematical sentences are used to state experimental facts is not dogmatic. Over and over again in *RFM*, he returned to question whether or not mathematics is an experimental science. (His assumption was that, if mathematical sentences are used to state facts, they state empirical facts. This assumption was revealed in V-15.) But his answer is always the same: Although mathematics depends upon facts, mathematical statements are not factual.

The reason for his not being able to identify empirical facts as mathematical truth conditions, and thus not being able to find any mathematical truth conditions, may well have been a fundamental position which he accepts unargued although not necessarily unconsciously. I believe that if we ask about the role of Wittgenstein's anti-platonism in his philosophy of mathematics and if we ask about the source of his anti-platonism in his philosophy, we can uncover some of the most fundamental of Wittgenstein's positions. I do believe that an insurmountable objection to accepting mathematical objects, in other words, an insurmountable objection to platonism, was the determining factor in Wittgenstein's philosophy of mathematics. It was a determining factor in the sense that any method of philosophizing or philosophical position which led to platonism would be rejected. If a philosopher had no insurmountable objection to platonism, I do not know why he would fight against the common belief that there are mathematical facts, reject the suggestion of our ways of speaking that there are mathematical facts, and develop a new method of philosophizing in order to handle traditional problems of philosophy without having to talk about truth conditions.

In this essay, I shall not criticize Wittgenstein for being anti-platonistic. I shall direct my efforts towards trying to present and evaluate his non-platonistic account. I suspect that most philosophers would agree that if a philosopher could develop a non-platonistic account, the non-platonistic account is to be preferred over any platonistic one. However, at times his resolve to be non-platonistic may seem dogmatic and uncritical in his philosophy of mathematics. But this anti-platonism may have a source outside his philosophy of mathematics.

20. So it is appropriate to digress on the source of Wittgenstein's anti-platonism. (This section is a digression in the sense that I shall be talking about his philosophy in general rather than about only his philosophy of mathematics.) His anti-platonism did not stem from an empiricism which holds that we can think about and get knowledge of only observable objects. In my opinion, his anti-platonism stemmed from his belief that no fact must be as it is. This belief in the contingency of all facts remained unchanged since his early thought in the *Tractatus*. In his *Tractatus*, Wittgenstein wrote that necessary truths and falsehoods cannot be statements of fact. As he put it in 4.462:

> Tautology and contradiction are not pictures of the reality. They present no possible state of affairs,

In 5.43 he wrote:

> But all propositions of logic say the same thing. That is nothing.

And in 5.1362, in a passage strikingly similar to the ones we cited from V-2 he wrote:

> ("A knows that p is the case" is senseless if p is a tautology.)

(It should be noted that in the *Tractatus* Wittgenstein held that there is only logical necessity in 6.37.) In addition in 5.5303 of his *Tractatus* he maintained that statements of identity are not statements of fact because they are necessarily true or necessarily false. I am sure that the above passages show that in the *Tractatus* Wittgenstein would not admit that there is a kind of fact which provides truth conditions for necessary truths.

It seems to me that his willingness to talk about the constituents of facts in the 2s section of his *Tractatus* indicates a willingness to talk about unobservable objects while doing philosophy, although I doubt that he would have admitted that these unobservable constituents of facts ever formed an unobservable fact. So even at the *Tractatus* stage of his thought empiricism was not a major concern of Wittgenstein. In Wittgenstein's later writings there are, to my knowledge, no claims that there are unobservable objects, no claims that there are no unobservable objects, and no claims that talk about unobservable objects are meaningless. I submit that in his later writings Wittgenstein was unconcerned with how statements of fact are connected with their truth conditions and the nature of these truth conditions. He was not concerned with whether or not descriptive terms in a sentence have observable referents. Nor was he concerned with whether or not factual statements have observable confirming or disconfirming conditions. I interpret Wittgenstein's famous claim in PI-43: "For a *large* class of cases – though not for all – in which we employ the word 'meaning' it can be defined thus: the meaning of a word is its use in the language" as expressing a lack of concern for the traditional empiricists' goals of establishing meaningfulness of sentences by finding observable referents for descriptive terms or finding observable confirming or disconfirming conditions. His statement in V-13 that propositions about physical objects and sense impressions do not need an analysis also reveals a lack of concern with the traditional problems of empiricists. So I conclude that in his later writings his anti-platonism does not come from an empiricist bias. And his indifference to empricism in his later writings was no reaction against his earlier work; he simply was never bothered by the problems of empiricism.

However, I must admit that Wittgenstein wrote in *RFM* as if the only facts that there are are empirical facts. In V-15 he claimed that only experiment can uncover facts. To my knowledge he never challenged this claim that only experiment can uncover facts. In III-4 he wrote:

'This could be imagined otherwise' or 'We can imagine the opposite too', ascribes the role of an empirical proposition to it.

So it seems that Wittgenstein identified being a contingent fact with being an empirical fact and being an empirical fact with being a fact. Thus I conclude that in *RFM*, as well as in his *Tractatus*, Wittgenstein believed that there were no necessary facts, i.e., no facts provided truth conditions for necessary truths.

But how does a belief that there are no necessary facts lead to anti-platonism? If there were timeless objects, as the mathematical objects of platonism would be, they would stand in unchanging arrangements. But objects, i.e., individuals and properties, in unchanging arrangements would constitute necessary facts. So a rejection of necessary facts requires a rejection of the mathematical objects of platonism.

My purpose in going into this digression on the source of Wittgenstein's anti-platonism in his philosophy of mathematics is to show that it was not based on a dogmatic, sophomoric rejection of abstract objects simply because they are abstract, or as Bernays said, outside concrete reality. I would like to suggest that his anti-platonism sprung from a *Weltanschauung* expressed in 6.41 of his *Tractatus*. In 6.41 he wrote:

The sense of the world must lie outside the world. In the world everything is as it is and happens as it does happen. *In* it there is no value – and if there were, it would be of no value.

If there is a value which is of value, it must lie outside all happening and being-so. For all happening and being-so is accidental.

What makes it non-accidental cannot lie *in* the world, for otherwise this would again be accidental.

I submit that in his *RFM*, as in his *Tractatus*, Wittgenstein held that the world is the totality of facts, only natural science can give us knowledge of these facts, these facts are all that can be known, and none of them has to be as they are. The major difference between his early and later philosophy is that in his later philosophy he had a new method of philosophizing which prevented him from doing the metaphysics behind his pessimistic *Weltanschauung* but which enabled him to provide a far more accurate account of the role of mathematical and other non-scientific language although the non-scientific language could never, strictly speaking, be used to express knowledge.

21. Let us return from this digression to a discussion of the three groups of questions on the philosophy of mathematics. Of course, if we give a non-platonistic answer to (Iii), we can skip down to the second group of questions on the modality of mathematical propositions. On the other hand if we give a platonistic answer, we face (Iiii). What (Iiii) asks is straightforward. Actually (Iiii) can be fully answered only by answering the questions in Group II. Also the great difficulties a philosopher faces in answering (Iiii) may well lead him to recant his platonistic position and hence his view that mathematical sentences have truth conditions. This recantation will bring him directly to the Second Group of questions. So without further comment, let us turn to (IIi).

Question (IIi) is designed to make those who have denied that mathematical sentences have truth conditions commit themselves on whether or not it is legitimate to talk about there being any mathematical truths or falsehoods. Of course, (IIi) is a genuinely open question for a philosopher who has denied that mathematical sentences have truth conditions, viz., who has denied that mathematical sentences are used to state facts in non-applied mathematics. One can look at the program of Logicists, such as Bertrand Russell, according to which they would try to express all existing mathematics as logical truths containing no unbound descriptive signs as an attempt to show how there could be mathematical truths, indeed necessary truths, without there being any mathematical facts. However, answering (IIi) is not easy for a non-platonist. A platonist need simply say that mathematical propositions are true if they correctly describe the mathematical facts; otherwise they are false. But the non-platonist faces the dilemma of either saying that mathematical statements are neither true nor false or else explaining how they can be true but yet true of nothing. One has to oppose ordinary usage of "true" too much to say that mathematical statements are neither true nor false. (One does not, however, oppose ordinary usage if he points out that the conditions for applying "true" to a mathematical statement are radically different from the conditions for applying "true" to an empirical statement. I should also add that to say that the conditions for applying "true" in mathematics are different from those in natural science is not necessarily to say that there are two senses of "true".) It is difficult, but a philosopher can come to grips with the other horn of the dilemma. However, a philosopher cannot come to grips with it by holding that mathematical statements are contingently true but yet have no truth conditions. It seems to me that for a statement to be contingently true it must correspond to a fact which does occur but which need not have occurred. I think that someone is misusing "true" if he says that a proposition, or a statement he makes, is contingently true because its truth is contingent upon a rule or convention saying that it is true. To say "S is true because S is called true" is odd. I suppose that a platonist could try to maintain that mathematical sentences are used to express contingent facts about the mathematical domain. But I do not see how timeless objects could stand in contingent arrangements. Thus if a philosopher answers (IIi) affirmatively he will very likely answer (IIii) by saying that mathematical truths are necessarily true. That is: a philosopher who answers (IIi) affirmatively will very likely go on to say that mathematical truths are necessary truths unless he holds that mathematical sentences are used to state empirical facts. If a philosopher holds an empirical interpretation of mathematical statements, he will have his difficulties with question (IIii) on the necessity of mathematical truths. But before we say anything about (IIii) let us try to attain an idea of how Wittgenstein answers (IIi).

22. Wittgenstein was tempted to say that what mathematical sentences are used to state or that which they express is neither true nor false. To say that what they are used to state has truth value very readily leads us to believe that they state facts. He revealed this wish in Appendix I-4 when he remarked that there is only a superficial similarity between what is expressed with

arithmetic sentences, such as $2 + 2 = 4$, and everything else that we call a proposition. But he did not oppose ordinary usage. Throughout his book he talked of mathematical propositions. At places he talked explicity of mathematical truth. In II-46 he asked himself: "When do we say that a mathematical proposition is true?" From the context it is clear that he did not expect the answer to be "never". In V-36 he remarked: "The proof of a proposition shews me what I am prepared to stake on its truth." So it is clear that Wittgenstein answered (IIi) affirmatively, i.e., he would talk about true and false in mathematics.

If I am right, then, in maintaining that a philosopher who does not give an empirical interpretation of mathematics but who holds that there are mathematical truths will answer (IIii) by saying that mathematical truths are necessarily true, Wittgenstein should have held that mathematical truths are necessarily true. And this he did. As a matter of fact, the central problem of *RFM* can be viewed as the problem of explaining an affirmative answer to (IIii). Recall that explaining an affirmative answer to (IIii) is to explain the nature of mathematical necessity and to explain the nature of mathematical necessity is to answer (IIiii). Early in *RFM*, at I-4, Wittgenstein asked himself: "But then what does the peculiar inexorability of mathematics consist in?" In the case of the problems of mathematical necessity we have a clear case of philosophical problems being problems about understanding our grammar if one, not unreasonably, regards the modality of a sentence as one of its grammatical features. So at least in this case, Wittgenstein was not inaccurate when he wrote in V-13: "What mathematical propositions do stand in need of is a clarification of their grammar." That is, he was not inaccurate if, as I submit, one of the central problems of *RFM* was to help us understand how mathematical sentences can be used to say what must be so.

23. Let us now consider the third group of questions. It is clear, I hope, that the third group of questions asks whether applied and non-applied mathematics differ, if at all, by their subject matter or modality. The intent of asking the third group is to provide some check on the answers to the first two groups. At

least a necessary condition for being an acceptable philosophy of mathematics is to be able to account for applied mathematics. I suppose that the only completely satisfactory check on one's philosophy of mathematics is to integrate it into a satisfactory philosophy. Of course, the third group of questions is also designed to answer philosophic doubts about how mathematics can be both pure and applied.

24. Wittgenstein's answers to the questions about applied mathematics are an important part of his philosophy of mathematics. One of Wittgenstein's most important positions was that mathematical sentences are meaningless without application. In IV-2 he wrote:

> one can form a mathematical proposition in a grammatically correct way without understanding its meaning.
> Now when does one understand it? – I believe when one can apply it.

In IV-2, he suggested that it is application which makes a game with signs into mathematics. (Presumably, mathematical sentences, which Wittgenstein calls "propositions", without application would be meaningless and hence their use would be a mere game with symbols.) A consequence of his position that there are no meaningful unapplied mathematical propositions is a hint that a proper understanding of applications would remove the antinomies. In V-8, Wittgenstein noted:

> What Russell's '∼f(f)' lacks above all is application and hence meaning.

If '∼f(f)' were meaningless we would not have to worry about any of its consequences. So there is no doubt that Wittgenstein thought that a philosopher of mathematics must pay attention to applied mathematics.

25. Can an attempt to answer the above three groups of questions be called doing the philosophy of mathematics? Of course, there is no doubt that dealing with the ontology, epistemology, and difference between pure and applied mathematics is philosophizing about mathematics. The only question is whether or not

a philosophy of mathematics should concern itself with more. In particular, should a philosophy of mathematics include analyses of crucial mathematical concepts and an attempt to provide mathematical foundations for mathematics? In section 27, I will point out why Wittgenstein thought that labor on the mathematical foundations of mathematics was philosophically irrelevant. But in this section let us see what Wittgenstein would have said about analyzing mathematical concepts.

In his *The Nature of Mathematics* (Routledge and Kegan Paul, 1933), in a chapter entitled "Tasks of a Philosophy of Mathematics" Max Black wrote the following.

> A philosophy of mathematics has two principal objects intimately connected with arithmetic and the theory of functions respectively: –
> (1) To elucidate and analyze the notion of 'integer' or 'natural number',
> (2) to alucidate the nature of the mathematical continuum.

For Black an elucidation or an analysis of a notion such as 'integer' is what he called a logical analysis of it. For him a logical analysis of a notion consists in showing how the motion should be connected with other notions so that with the analyzed notion we can say everything we said with it before except those things which led to contradictions or which were philosophically (logically) misleading. For instance, a logical analysis of the notion of class or set should show how we can do all of set theory except for those parts which led to the antinomies. In his chapter on logical analysis, Black cited Russell's analysis of the notion of definite descriptions as a good example of logical analysis. In his analysis of definite descriptions, Russell showed that once we understand how the notion of a definite description is related to other notions such as that of existential quantification, we shall no longer be misled to believe that there must always be some kind of referent for a definite description. Frege's attempt to define "integer" is also an example of logical analysis. In *RFM*, one finds no logical analyses.

26. I think that Wittgenstein was unconcerned with logical analyses because, as we noted in section 8, he was concerned with

understanding mathematics as it is now done. A logical analysis of current mathematics will alter the mathematics now done in the following ways. It will improve it mathematically by removing contradictions, alter its forms of expression by elimination of allegedly philosophically misleading expressions, and it will alter the structure of present mathematics by containing connections not contained in present mathematics. For instance, the arithmetization of analysis was an alteration of the present structure of mathematics. For Wittgenstein such an alteration of mathematics would be philosophically useless and, I suspect worse than useless since looking at the new improved mathematics would not be a study of mathematics as it is now done. Also it is possible that a logical analyst could so seriously alter the ways of expression used in mathematics that the IMPROVED mathematics is no longer mathematics. (In Chapter IV, we will find that Wittgenstein argued that Logicists had so seriously altered mathematics that they no longer had mathematics after their analyses.)

But it is more important to see why Wittgenstein would regard such analyses as philosophically useless since the efforts of mathematicians may actually lead to the kind of mathematics which a logical analyst thinks that there ought to be. In case of such a development, the job of a philosopher would be to understand the new IMPROVED mathematics. For one, the mere removal of contradictions leaves unanswered the philosophically significant question: Why remove contradictions? In my opinion the question: "Why remove contradictions?" is philosophically significant because an answer to it presupposes positions on the ontology of mathematics, the epistemology of mathematics, or positions on the nature of applied mathematics. For instance, one philosopher may claim that contradictions are inadmissible because contradictions cannot correctly describe the mathematical domain. Another philosopher may hold that contradictions are disastrous because the only way through which we can come to know most mathematical propositions is through logical derivation from previously proved propoistions and every proposition can be derived from a contradiction. Also some philosophers may object to contradictions primarily because they can have no application. In

RFM, we will find Wittgenstein trying over and over again to get philosophers and mathematicians to ask themselves why they abhor contradictions. A second reason why it is philosophically useless to give a logical analysis of a mathematical notion is that such an analysis will exhibit connections not recognized in mathematics as it is presently done. But the philosophically interesting question is: In mathematics do we discover or invent connections? Merely exhibiting a connection is irrelevant for the philosphical question. Finally, removal of allegedly philosophically misleading expressions is useless and may actually be pernicious. Such elimination is useless because it is not the expressions themselves which are defective. Their only defect is our defect of putting some incorrect or unclear philosophical interpretation of their use upon them. In *PI*-194 and in I-125 Wittgenstein wrote:

> We pay attention to the expressions we use concerning these things: we do not understand them, however, but misinterpret them. When we do philosophy we are like savages, primitive people, who hear the expressions of civilized men, put a false interpretation on them, and then draw queer conclusions from it.

The proper thing to do would be to get a clear understanding of their use, – become civilized –, so that we no longer give the incorrect or unclear interpretation of their use. Would it not be foolish to give up the expressions of civilized men for new barbarisms? Also why should we think that some ways of expressing ourselves are philosophically better? Should we not ask ourselves whether or not we are assuming that some ways of speaking present a philosophically correct picture of reality? And if so, should we not justify such an assumption? Wittgenstein himself believed that many logical analysts incorrectly assumed that we can get a way of speaking, which will lead us to the right picture of reality, if we use the notation of mathematical logic. He remarked in IV-48:

> 'Mathematical logic' has completely deformed the thinking of mathematicians and of philosophers by setting up a superficial interpretation of the forms of our everyday language as an analysis of the structure of facts.

(In chapter IV, section 17, there are more comments on this remark from IV-48.)

27. So one should not expect analysis of mathematical notions in *RFM*. But what about work in areas such as improving the rigour of mathematical proofs and trying to establish the consistency of portions of mathematics? Should not such investigations be part of a philosophy of mathematics? I would say that a person is working on the mathematical foundations of mathematics if he is investigating the cogency of proofs, the interconnections of various branches of mathematics, attempting to prove the consistency, or discussing the completeness, categoricity, or independence of axioms for all or part of what is called mathematics; and he can accomplish, or believes that he can, his ends by means of mathematical techniques. I think that passages that I have cited so far show that Wittgenstein was not at all interested in what I have just called work on the mathematical foundations of mathematics. That is to say that he is not interested in these investigations except insofar as they provide more mathematics to philosophize about.

So a reader of *RFM* will not have to struggle with problems in mathematical logic. Of course, his refusal to become entangled in problems of logical analysis and logic eliminates labor in coming to understand his position only if it is compensated by his coming to grips with the philosophical issues in a clear way. It may well turn out that his non-technical way of approaching the philosophic issues is more demanding than technical problems and less satisfying intellectually because it does not have the clarity of mathematical method. Unfortunately, Wittgenstein's approach to these questions about the philosophic foundations of mathematics is difficult to follow. Hence, it may be helpful if I give my interpretation of his method of philosophizing about mathematics.

28. How should one try to answer these questions on the philosophic foundations of mathematics? There are at least two ways to approach these questions. One way is direct and rather naive while the other is devious and more sophisticated. But before I say anything specific about either of these two approaches, I shall discuss what they have in common. This discussion of what

they have in common is partly an elaboration on what we have already discussed in section 9 but I believe that further discussion will be helpful towards placing Wittgenstein's method in perspective.

One is not going to approach these questions at all unless he has some problem about, – some wonder about –, the ontology, epistemology, and applications of mathematics. The questions arise when one has certain beliefs, or biases if you will, about the ontology, etc., but for some reason realizes that there are difficulties in accepting these beliefs. The man questions, then, or wonders, how these beliefs can be reconciled with the difficulties. For instance a philosopher may have the belief that mathematical statements are about objects. Possibly, the similarity between "Bill is the son of John and Mary" and "Five is the sum of two and three" led him to such a belief. But nonetheless he has it. Yet, for the kind of reason we noted in our discussion of platonism in section 17 he realizes that mathematical objects would be non-ordinary. However, he has some doubts about the existence of non-ordinary objects, let us say, because of their timeless character. Hence, he has the problem of reconciling his belief, – his expectation –, that his mathematical statements are about facts with the apparent possibility that there are no facts for them to be about. Another philosopher may believe that mathematical propositions are necessarily true but wonder how this could be so if all knowledge is gained from experience which only shows us what is the case and never what must be the case. A third example of having a problem about the philosophy of mathematics would occur in the case of a man who believed that in applications of mathematics we regarded ordinary objects as mathematical objects. Such a man would wonder how mathematics could be applied without distorting what we applied it to. So in my opinion any philosopher who approaches these questions approaches them because he has a preconceived idea of how they should be answered; he has also, however, some doubts about the correctness of his preconceived answers, and he approaches these questions to reassess his preconceived answers.

29. How would a philosopher go about reassessing his pre-

conceived answers? One does not have to use the method I will suggest but I think that it is the most reasonable one and the most frequently used method. The method I wish to present will be more intelligible if I first illustrate it with an example. (I should also add that I am still talking about what is common to the naive and sophisticated approaches.)

Assume that a philosopher has the preconceived answer that mathematical propositions are about non-ordinary objects and that he wants to reconcile this answer with his skepticism about the existence of non-ordinary objects. How can he proceed? First, he would try to find a kind of proposition which he could regard as a proposition about objects which have most, if not all, the features which he attributes to mathematical objects. For instance, he may pick propositions from finite arithmetic or he may pick propositions about small finite classes. By consideration of this kind of proposition, he would try to get a clear idea of what it is like for there to be non-ordinary objects and to convince himself that there are such. In his efforts to get a clear idea of what it is like for there to be non-ordinary objects he may have to do some metaphysics to place these non-ordinary objects amongst the ordinary objects. To convince himself that there can be these non-ordinary objects, he may have to revise some of the metaphysical beliefs which originally led him to be skeptical about the existence of non-ordinary objects. For instance, he may have to give up a belief that only observable individuals can exist. If he is lucky enough to discover a kind of proposition which he is certain is about non-ordinary objects, the immediate effects of his reassessment will be in his metaphysical beliefs. To find a place for non-ordinary objects in his scheme of things, he will reluctantly, but relentlessly, eliminate or alter beliefs which are inconsistent with there being non-ordinary objects. Of course, if he cannot find such a kind of proposition, if he cannot find a place for non-ordinary objects, or if he is simply convinced of his beliefs which are inconsistent with there being non-ordinary objects, the result of his reassessment may be either a rejection of the original answer that there are mathematical objects or a continuation of the tension of an unsolved philosophical problem.

But to continue our example, let us assume that our philosopher has considered propositions about small finite classes, has obtained a *clear* idea of the place of such classes in his scheme of things, and has convinced himself that classes are non-ordinary objects which exist. Despite all of this, he has only begun to philosophize about mathematics; he has only selected his paradigm of a mathematical proposition. His next task is to show how all mathematical propositions can be construed as being the same kind of proposition as his paradigm. (I shall explain in section 31 why I use the vague term "construe". In our example, this first amounts to trying to show how all mathematical propositions can be re-expressed as propositions about small finite classes. When our hypothetical philosopher realizes that he cannot re-express mathematics as propositions about small finite classes he, unless he wants to give up, will convince himself of the metaphysical belief that for purposes of ontology all finite classes are the same. He probably will not have too much difficulty in admitting any finite class. Soon our hypothetical philosopher will realize that he needs to admit infinite classes to re-express all mathematical propositions as propositions about classes. He must then not only add an axiom of infinity for mathematical purposes, he must also philosophically reconcile himself to the position that if he admits finite classes there is no reason why he should hesitate to accept infinite ones. After he has accepted infinite classes, the antinomies require him to admit that his concept of what it is like for there to be a class was not as clear as he had originally believed. As a matter of fact until he can get a consistency proof of his class theory revised to avoid the antinomies, he must admit that he does not have a perfectly clear idea of what it is for there to be an infinite class. (And if all classes are the same kind of thing, he must even admit that he does not have a perfectly clear idea of what it is for there to be a small finite class.) Still, once he has construed mathematical propositions as being the same kind of propositions as his paradigm kind, his philosophizing about mathematics is not yet finished. His next task is to see whether or not he can give adequate answers to the questions in Groups II and III. (By construing mathematics as class theory, he has already answered the questions of the first group on mathematical ontology.)

30. My purpose in sketching this set-theoretic way of approaching these questions on the philosophical foundations of mathematics was to illustrate what I think is the most reasonable and most common approach to these questions – and, not incidentally, the approach taken by Wittgenstein in *RFM*. But let me emphasize that Wittgenstein did not try to construe mathematical propositions as being about sets or classes, be they large or small. The approach is for one to try to find a kind of proposition, – a paradigm set of propositions –, such that if mathematical propositions were like this kind of proposition, then one could see how his original answers about the ontology, epistemology, or applications of mathematics could, with or without modification, be true. Then he would try to show that mathematical propositions are sufficiently like the paradigm kind of propositions. If by this method he reached a fairly satisfactory solution to one of the problems, e.g., problems about ontology, he would then reconsider his paradigms to see whether or not they made it clear how the other philosophic questions about mathematics can be satisfactorily answered. If reconsideration of one's paradigms did not provide satisfactory answers to the other questions, he could either look for a new paradigm kind of proposition or lower his standards on what is a satisfactory answer. However, in the case of Wittgenstein (as well as in the case of those who believe that propositions about classes are suitable paradigms), he had a fairly good idea that his respective paradigms were paradigms for his original answers about ontology, epistemology, and applications. This is so because philosophers rarely restrict their attention to one issue at a time. However, it is not misleading to regard philosophers as working in the systematic way that I suggested.

31. Before I turn to a comparison of what I called the direct and devious approaches, I must say something about my use of the words "construe" and "sufficiently like". What is it to construe a mathematical proposition as being of the same kind as a paradigm kind of proposition? What is it to show that a mathematical proposition is sufficiently like a paradigm proposition? For in-

stance, what is it to show that propositions of mathematics are sufficiently like propositions about classes or what is it to construe them as propositions about classes? I think that it is a mistake to lay down precise criteria for saying that one has construed mathematical propositions as being identical with or sufficiently like a paradigm kind of proposition. For one, we cannot identify mathematical propositions with anything but mathematical propositions. We should take seriously Butler's maxim: "A thing is what it is and nothing else" with which G. E. Moore began his *Principia Ethica.* Once we can no longer demand identity, the question of what degree of similarity we should accept can be a subject of endless dispute unless we have some specific purpose for saying that the mathematical propositions are (are not) sufficiently like a paradigm kind of proposition. Our purpose in saying that a mathematical proposition is like a paradigm proposition is to transfer what we say as philosophers about the paradigm back to the mathematical proposition. So the construing will occur as part of a philosophical argument. Hence, before we can start to determine whether or not mathematical propositions are like a philosopher's paradigm we should consider his paradigm, what remarks about his paradigm he wants to transfer to mathematics, and his justification, if any, for transferring the remarks. I think that trying to lay down criteria on what is a satisfactory analysis of mathematical propositions, where "analysis" means construe them as another kind of proposition, would be tantamount to criticizing arguments before they are presented. In general, and in philosophy in particular, it is always best to look at the arguments rather than some general considerations on what can and cannot be done. Still, I think that the following two generalizations can be made. To construe proposition S as being of the same kind as proposition Q is to regard the subject matter of S, the ways of coming to know S, and the uses of S as being of the same kind as those of Q. If proposition S is sufficiently like proposition Q, then we could have originally done with S what we now do with Q without ever having been familiar with Q.

32. We can now turn to contrasting the direct and the devious approaches. But the effort spent in discussing what they have in

common was not wasted since it shows how much Wittgenstein's approach to the philosophy of mathematics has in common with more standard approaches. A direct and naive approach to these questions involves developing an ontology and epistemology. What I call the devious and sophisticated approach involves trying to answer these questions without developing an ontology or an epistemology. To illustrate the difference let us return to the example, in section 29, of a philosopher who tries to develop a philosophy of mathematics by talking about set-theory. Recall that at one stage this philosopher had to develop a clear idea of what it is like for there can be classes and to convince himself that classes exist. This would be doing ontology. Later on he would have to become clear about how humans can attain knowledge of necessary truths about classes. This would, of course, be doing epistemology. It is the doing of the ontology and epistemology which can be done directly and naively or deviously and sophisticatedly. It is done directly and naively if one develops a theory about the kinds of objects that there are and a theory on how they can be known. This theory would be such that it would enable the philosopher to deduce that if mathematical propositions are sufficiently like the paradigm propositions, then answers quite a bit like his original answers to the questions would be correct; he could, at the same time, deduce that his doubts about the correctness of his original answers could, with minor concessions to them, be dismissed. In our example, a philosopher may develop a theory on the place of classes amongst the things in the universe which would enable him to deduce his original answer that mathematical propositions are about some non-ordinary objects necessarily connected with one another, while deducing the incorrectness of his original doubt that we can have knowledge only about observable objects. In effect, a philosopher who takes the direct and naive approach treats philosophical problems as scientific problems on how to explain some phenomenon; he tries to remove wonder by showing that what was wondered about is exactly what is to be expected according to the correct theory. (Recall the quotation from Wittgenstein's *Blue Book*, in section 12, where he blamed use of the method of science in philosophy as the source of metaphysics.) I call such an approach direct because such an approach is what seems to be called for by the questions on mathematical ontology and epistemology. I call it naive because it is assumed that we can develop metaphysical theories. Even if the philosopher calls his theory *meaningless,* it is at least assumed that a presentation of it and use of it will solve the philosophical problems.

However, a sophisticated philosopher, – sophisticated about the difficulties of metaphysics –, who is convinced of the impossibility of metaphysics may try to accomplish the same goal as the naive philosopher. The common goal is the removal of wonder. But he will try to attain it in a devious manner. He may do everything the same as the naive philosopher except for the stating of the metaphysics, i.e., stating his ontology and epistemology. For instance, a philosopher taking the devious approach may use set theory in the following way. He would take a standard set-theory, such as von Neumann's, and then by talking about it would try to suppress any wonder about mathematical objects, epistemology, and applications. But he would take care to suppress wonder without stating any metaphysical theory. Most likely he would answer questions about the existence, nature, and connections of mathematical objects by discussing examples of classes and by pointing out how classes are talked about in set-theory. For instance, he may convince someone of the existence of classes by convincing him that a pair of shoes is different from each shoe. But he would stop after convincing the person that a pair of shoes is not a shoe; he would not state that classes exist. He may convince someone that infinite classes exist by pointing out that in set theory an infinite set is said to exist; but he himself will not explicitly assert that an infinite set exists. I am not suggesting that any philosopher who took the devious approach would try to philosophize about mathematics by talking about set-theory; I am sure that he would not. I simply want to show that an essential feature of the devious approach is not talking about ordinary language as opposed to a formal language. As I see it, the essential difference between the direct and devious approaches is that the first uses philosophical propositions while the second will not utter them. The direct approach removes wonder by explaining why the thing wondered at occurs. The devious approach tries to describe, – gets us to look at –, the thing wondered about from many points of view until it no longer seems something to be wondered about.

33. There is no doubt that in his *Blue Book, PI,* and *RFM* Wittgenstein took the devious approach. This is why he wrote in *PI*-133:

> The real discovery is the one that makes me capable of stopping doing philosophy when I want to. – The one that gives philosophy peace, so that it is no longer tormented by questions which bring *itself* in question. – Instead, we now demonstrate a method, by examples, and the series of examples can be broken off. – Problems are solved (difficulties eliminated, not a *single* problem.
>
> There is not a philosophical method, though there are indeed methods, like different therapies.

I take him to mean that once a philosopher has removed wonder about one issue he should be able to stop philosophizing about that issue. He can stop if the removal of wonder did not require the development of a metaphysical theory which itself requires justification. If his removal of wonder required a metaphysical theory, his removal of wonder is contingent upon how well his metaphysical theory solves other problems since his metaphysical theory will at least connect several problems, if it does not unite all philosophical problems into the single problem of finding the correct metaphyhics. But worst of all: If the removal of his wonder requires the development of a metaphysical theory, the notorious difficulties in metaphysics may lead the philosopher to despair of ever removing the wonder. So the gist of what Wittgenstein meant by saying that a philosopher should be able to stop philosophizing is that after a philosopher solves a particular problem he should not have to worry about the ramifications of a metaphysical theory devised to solve the problem. I do not, despite the fact that he calls philosophy a therapy, believe that Wittgenstein thought that the goal of philosophy was to stop men from philosophizing, where by "philosophizing" I mean the removal of wonder not removable by the explanations of natural science.

34. What is the correct approach? The direct approach is the natural approach. It is natural to answer direct questions directly. In particular, it is natural to answer ontological and epistemological questions by saying something such as: "There are not these kinds of objects because there can only be these other kinds" or "This kind of thing can be known because there are such and such objects and the mind has such and such capabilities." I think that the only reason that one has for taking the devious approach is despair over the possibility of metaphysics. But I think that it is a mistake to try to determine which approach is correct. A philisopher's approach is part of his argument and I believe it is best not to evaluate an argument until it is completely laid out. Without developing any metaphysics, Wittgenstein may remove our wonder about how it can seem so certain that mathematical truths are truths about objects in necessary connections while at the same time it can seem so reasonable to deny platonism. But we cannot tell until we see his arguments. Even if we do discover a metaphysical theory behind Wittgenstein's philosophyzing about mathematics, this does not mean that his solutions are defective. It means only that his declared method of philosophizing is not adequate for solving philosophical problems. It may even turn out that his implicit metaphysics, if he had one, when developed is the correct metaphysics. I think the fact, if it is a fact, that Wittgenstein's approach is inadequate is unimportant. The important thing is whether or not his solutions to particular philosophical problems remove wonder about one area without producing wonder about another. So, since the important things to study are Wittgenstein's actual philosophizing, we shall turn to his paradigms which exhibit his original answers to questions about mathematical objects and necessity. But before we consider his paradigms in Chapter II, let us return to a consideration of the role of "I must have miscalculated."

35. The role of a sentence is its function, its implications, and criteria for correct use. Questions about the ontology, epistemology, and even applications of mathematics arise when we ask about the function, implications, and criteria of "I must have miscalculated." For instance, assume that one tries to solve: $x + y$

$= 13$ and $x = 3y$. But somehow he concludes that $y = 2.25$ and $x = 10.75$. (Very likely his error would be due to careless writing which made a 3 look like a 2.) However, it might have occurred, a check of his result shows that $3 \times (2.25)$ is not 10.75. He may then say "I must have miscalculated." What is the function of, or purpose in, admitting that he miscalculated? Although, one is saying that his answer is unacceptable, he is clearly saying more than that. A simple claim that it is unacceptable would suffice to say that it is unacceptable. The admission of miscalculation, when an unacceptable result is obtained, reflects a belief of our culture that not everything anyone happens to calculate can be considered correct. It also reveals the belief that the correct mathematical answers can be arrived at by correct thinking, or at least the belief that incorrect ones are arrived at by incorrect thinking. Once we try to understand the function of "I must have miscalculated", we are trying to understand why we believe that there is a mathematical truth to be known by some correct method. We are also trying to understand what this truth is and what this way of knowing is. Such an investigation will surely involve asking the questions in Groups I and II. If our miscalculation occurred in applied mathematics, we could go on to consider whether or not our mistake was about the facts to which the mathematics was applied or in the mathematics itself. Such a line of thought may lead us to try to find the difference between applied and non-applied mathematics. And this is the intent of the third Group of questions. An implication of "I must have miscalculated" is that at some stage in one's work one did something called miscalculating. To try to understand what miscalculating is will certainly involve asking the questions about mathematical ontology and epistemology. For instance, what is one thinking incorrectly about when he miscalculates? Or, how can one even think when he miscalculates since if mathematical truths are necessary truths mathematical falsehoods would seem to be unthinkable absurdities. Another implication is that once the miscalculation occurred further mistakes necessarily followed. A question about this implication would be: What is mathematical thinking like so that further results are determined by a given result? Of course, if we ask about the criteria for correctly using "I must have miscalculated" we are immediately concerned with mathematical epistemology. So since an attempt to understand the role of "I must have miscalculated" leads us to ask the questions which provide a philosophic foundation of mathematics, it is not rash to call such an attempt a key to the foundations, – philosophic foundations –, of mathematics.

36. This may have been the key that Wittgenstein actually used. He frequently reminded himself that calculations are not experiments on what results we or others happen to get. This shows that he was trying to understand the belief that mathematical truth is objective but yet not descriptive of behavior. We just saw that a function of "I must have miscalculated" is to express this belief. Over and over again, he tried to become clear about what it meant to say that one step in a calculation proceeds necessarily from another. As a matter of fact *RFM* begins with such a problem. We have just seen that such an issue arises when we try to understand one of the implications of "I must have miscalculated." In passages such as I-112 and I-116 Wittgenstein showed that he was very concerned with understanding how we can think a mathematical falsehood. In IV-28, he questioned the use of *reductio ad absurdum* proofs because use of them requires us to assume a mathematical falsehood. In V-19, he went so far as to claim that his main difficulty with Godel's Incompleteness Proof lay in trying to understand how we could assume a mathematical falsehood. We have just seen that we would have to understand how we can think mathematical falsehoods if we are to understand the role of "I must have miscalculated." And as the cited passages show, Wittgenstein was concerned with this issue of how we can think what is mathematically false. So, Wittgenstein may have actually used this key in his philosophy of mathematics although I must admit that his remarks do not contain any explicit study of "I must have miscalculated."

II

STRICT FINITISM

> **The proposition 'a=a', 'p ⊃ p', "The word 'Bismarck' has 8 letters", "There is no such thing as reddish-green", are all obvious and are propositions about essence: what have they in common? They are evidently each of a different kind and differently used. The last but one is most like an empirical proposition. And it can understandably be called a synthetic *a priori* proposition.**
>
> **It can be said: unless you put the series of numbers and the series of letters side by side, you cannot know how many letters the word has. (III-39)**

1. In the preceding Chapter, I suggested that a philosopher who believed that mathematical sentences were used to express necessary truths but could not be used in non-applied mathematics to state facts, may begin to answer the questions about the philosophic foundations of mathematics in the following way. Very likely he would be motivated to ask the questions because he realizes that someone can make a good case for platonism. In order to vindicate, and possibly modify, his non-platonistic position, he would select a kind of sentence with the following properties. He could say about a typical sentence of this kind that it is used to state a necessary truth which is not a statement of fact. (The kind of sentence which he picks is his paradigm kind of sentence while the particular sentence which he refers to is his paradigm sentence.) Under the assumption that mathematical propositions are expressed with sentences of his paradigm kind, he will try to give satisfactory non-platonistic answers to the three groups of questions on the philosophic foundations of mathematics. "Satisfac-

tory" means that he can give clear answers to all of the questions and his answers do not have undesirable philosophic consequences. Of course, the paradigm kind of sentence should be sufficiently like the sentences used to make mathematical statements so that the philosophy developed by talking about the paradigm kind of sentence can be regarded as a philosophy of mathematics.

2. In this Chapter, my primary aim is to develop a philosophy of mathematics which originates from an attempt to preserve the belief that mathematical statements assert necessary truths but are not statements of fact. (I should repeat that I mean by "statement of fact" a statement with truth conditions.) This philosophy will be developed by taking "The word 'Bismarck' has 8 letters" as a paradigm of its paradigm kind of sentence. I shall call this philosophy of mathematics: strict finitism. (I follow Kreisel who called Wittgenstein's philosophy of mathematics strict finitism in his *"Wittgenstein's Remarks on the Foundations of Mathematics"* in the August, 1958 issue of the *British Journal of the Philosophy of Science*. And I believe that I am developing Wittgenstein's philosophy of mathematics in this chapter. On p. 505 of *BP*, Dummett reports that Hao Wang suggested calling this philosophy "anthropologism". I prefer "strict finitism".) However, in this Chapter, my primary aim will not be to argue that Wittgenstein held the philosophy developed here, although I shall cite passages which suggest that I am developing insights of Wittgenstein. In Chapter III, I shall be primarily concerned with supporting my interpretation of Wittgenstein as a strict finitist. I should also mention that, in this Chapter, I shall not discuss the legitimacy of assuming that mathematical sentences are sufficiently like the paradigms of strict finitists so that strict finitism can be regarded as a philosophy of all of mathematics. We shall discuss the issue of the comprehensiveness of strict finitism in the last chapter when we criticize Wittgenstein's philosophy. But here our aim is to understand strict finitism. To gain this understanding we shall consider "The word 'Bismarck' has 8 letters" and other sentences of this kind.

3. Let us first rewrite "The word 'Bismarck' has 8 letters" as

"The number of letters in 'Bismarck' is 8." Let us call the rewritten version B. I rewrite the sentence because in the rewritten version it sounds more like a mathematical statement expressing the value of a function, and I do want to suggest that B is very much like a mathematical sentence. Also in the rewritten version the "is" is naturally read as the "is" of identity. In the original version one could easily read the "is" as attributing the property of having 8 letters to the word "Bismarck". I want the paradigm kind of sentence to contain the "is" of identity, i.e., be equations, because if the "is" of identity is not the only copula in mathematics it is certainly the most frequent.

After we have rewritten our original sentence as B, we shall convince ourselves that B is necessarily true. Or, I could just as well have said that we shall prove B. In this essay, by "proving a sentence P" I mean the process of establishing that what P is used to assert must be true. Then we shall analyze what we have gone through when we proved B. But the steps of the analysis will be illustrated by reference to a proof of another strict finitist paradigm considered by Wittgenstein in III-50 and 51. The second paradigm is: The reverse of the letters in OVER is REVO. After this analysis, we shall develop the philosophy of strict finitism by answering the three groups of questions. The answers will be illustrated by reference to yet another paradigm sentence. The third strict finitistic paradigm is: The sum of two and three is five. The third paradigm will show that strict finitism can at least construe some arithmetic sentences as being of their paradigm kind. Let us now turn to proving B.

4. Is it true that the number of letters in "Bismarck" is 8 and is it necessarily true? Of course, it is true. Still, taking a close look at how we establish that it is true is helpful in establishing that it is necessarily true. How do we establish that B is true? Again the answer is easy: We count the letters in a token of the type "Bismarck". (I follow what I believe is standard philosophical usage in calling an instance of a word actually written down a TOKEN of the word and the word which we say the instance is a token of the TYPE.) Let us take a closer look at this counting. One way of doing this counting is to do it in three stages. See Picture 1.

```
Stage 1:  Bismarck

Stage 2:  B i s m a r c k
          | | | | | | | |

Stage 3:  B i s m a r c k
          | | | | | | | |
          1 2 3 4 5 6 7 8
```

The first stage is the production of a typical token of "Bismarck". The second stage is an application of the operation: taking the number of, to the typical token of "Bismarck". In the third stage, we apply a numerical concept to the result of having applied the operation: taking the number of. (In effect, all counting of elements in configurations is like this; only rarely so simple-minded. If we would have counted aloud, we would have visualized a token of "Bismarck" or looked at an actual token, made eight sounds while correlating each sound with a letter and a number word. For instance, saying the word "four" is correlating the sound we make when we say "four" with the number word "four".) We could say that what was done in Stages 1 through 3 was an experiment to determine how many letters the word "Bismarck" contains. If we call it an experiment, we can call Picture 1, and especially Stage 3, a picture of how such an experiment would almost always come out. As a result of this experiment we possess as a contingent empirical truth: "The number of letters in 'Bismarck' is 8."

5. But we have not yet proved B. Or better, we have not yet shown that we can transform the empirical "The number of letters in 'Bismarck' IS 8" into the non-empirical "The number of letters in 'Bismarck' MUST BE 8." In II-30 Wittgenstein wrote: "A proof leads me to say: this MUST be like this." In reading Wittgenstein it is helpful to interpret "proving" and "calculating" as naming processes by which we determine that what a sentence says must be true in addition to merely being true. His "calculate" does not refer to the performance of arithmetic manipulations. I would go so far as to say that in *RFM* he uses "prove" and "calculate" synonymously. Of course, the experiment itself cannot be the

proof or calculation since it tells us only what IS the case. However, the experiment will play an essential role in the proof or calculation. In V-15 he remarked: "A calculation – it might be said – is perhaps part of the technique of an experiment, but it is by itself not an experiment." (That calculating is not experimenting is one of the few positions that Wittgenstein argues for in what could be called the standard way philosophers argue. For instance, see II-65 to II-76 and V-15.) So the problem of transforming the experimentally justified B into a proved proposition remains.

6. Picture 1 already is part of the proof; it is the proof pattern. In V-15 Wittgenstein remarked: "the calculation is the picture of an experiment as it practically always turns out". We just noted that Picture 1 pictures how an experiment to determine the number of words in "Bismarck" would almost always turn out. However, we cannot really say that Picture 1 by itself is the proof or calculation. By itself it is only a picture of the technique which we used to establish B experimentally. We must make use of or look at this pattern in a certain way in order to have a proof of B. In II-11, Wittgenstein explicitly claimed: "the pattern is not the proof". And in II-22 he wrote:

> The proof (the pattern of the proof) shews us the result of a procedure (the construction); and we are convinced that a procedure regulated in *this* way always leads to this configuration.

In II-24 he added: "The proof must be our model, our picture, of how these operations have a result." I should point out that in II-22 and II-24, Wittgenstein was writing about proving metalinguistic propositions of the form: This sequence of formulae from axiom system A is a formal proof of formula F. We shall see in sections 92 and 118 it is not inappropriate to take these remarks about meta-mathematical proofs as applying to any *strict finitistic proof*. So, although Wittgenstein sometimes wrote as if Picture 1 itself could be a proof, I submit that he really meant that pictures like Picture 1 were patterns for proofs. To get a proof from a pattern certain mental conditions must obtain. For instance, people must be convinced that the picture of what happens when an operation is applied to a figure is suitable as a model of what it is like to apply this operation to such figures. Let us now go on to develop a proof of B from Picture 1.

7. Call the typical token of "Bismarck" in Picture 1 X. (If one wants to be pedantic about the use of "token", call the token of "Bismarck" in stage 3 X. However, I intended the stages of Picture 1 to represent various stages of dealing with a single token of "Bismarck". Stage 3 shows us that it is empirically true that the number of letters in X is 8. But inspection of Picture 1 establishes more than this empirical truth. From Stage 3 of Picture 1 we can read-off that the number of letters in X must be 8. (I use read-off because Wittgenstein used "read-off" in this way. See *RFM* II-24.) Why do I claim that we can read-off that X must have 8 letters? It is inconceivable that X could be the token that it is and have more or less than 8 letters. If I am wrong about the inconceivability of X being a "Bismarck" token and having more or less than 8 letters, the simplest way to refute me is to draw another number of strokes, say 7 or 9, and picture how they can be put in one-to-one correspondence with the letters of X. Although you may not put it this way, you would agree with Wittgenstein's remark in I-45: "A demon has cast a spell around this position", e.g., having 9 letters, "and excluded it from our space."

At first glance, before one counts the letters in X and before one considers whether or not there can be more or less letters in X, one may think that X need not have 8 letters. Similarly, one may think that X need not be printed in the color in which it is printed. However, the facts are that we can think of X being a "Bismarck" token even if it is another color while we cannot think of it being one with more letters or less letters than 8. Similarly, at first glance one may not think that a red cube must have 12 edges to be a cube. Yet reflection established that it is inconceivable that this particular red cube be a cube without its 12 edges; although we can easily think of it being blue. In the case of the cube, reflection did not show us only that it is inconceivable that a cube have 12 edges and not have 12 edges. Reflection showed us that the particular cube could not be the figure that it is without 12 edges. Similarly, I am not merely claiming that it is inconceivable that X has 8 letters and X does not have 8 letters. I am making the much stronger claim that it is inconceivable that X has any amount of letters other than 8. (I shall argue in section 10 that "We cannot conveive of a token of 'Bismarck' having more or less than 8 letters" is not equivalent to "We cannot conceive of a token of 'Bismarck' having both 8 letters and simultaneously not having 8 letters.")

However, from observation of what IS the case with X, we can read-off a general necessary truth in addition to the singular necessary truth that X must have 8 letters. Since X is a typical token of "Bismarck", observation of and reflection on Stage 3 enables us to read-off, or intuitively induce, as a necessary truth the general statement: For any X if X is a token of "Bismarck", X has 8 letters. Or, since this general statement is equivalent to B, we can say that observation of and reflection on Stage 3 has resulted in a proof of B.

8. We can look at what we have done in establishing B as necessarily true in another, and I think illuminating, way. We could say that on the basis of what we see in Stage 3, we establish an implicatory link between the concepts "being a 'Bismarck' configuration" and "containing 8 letters". In suggesting this I follow Wittgenstein's suggestion in V-46 where he wrote: "The mathematical Must is only another expression of the fact that mathematics forms concepts." In V-49 he specified what he meant by forming a new concept when he wrote:

> "To give a new concept" can only mean to introduce a new employment of a concept, a new practice.

We could say that in the proof of B we legitimatized the practice of using "having 8 letters" as a necessary condition for applying the concept "Bismarck – configuration". Thus following Wittgenstein's specification of what it is to introduce a new concept, we could say that in the proof of B we establish a new concept of "Bismarck".

I would not object to a claim that in mathematical proofs we establish new conceptual connections, and hence new concepts, and that mathematical sentences are used to state what connections we have established. However, Wittgenstein hesitated to make such a claim. In V-33 he mused:

> Is a new conceptual connexion a new concept? And does mathematics create conceptual connexions?
>
> The word "concept" is by far too vague.
>
> Mathematics teaches us to operate with concepts in a new way. And hence it can be said to change the way we work with concepts.

In V-49, he again commented that "concept" is a vague concept. I attribute Wittgenstein's hesitation about making such a claim to his approach to philosophy. Recall from our discussion in Chapter I, section 32, that he did not want to burden himself with a philosophical theory in order to understand a non-philosophical practice. In this case, he did not want to burden himself with a theory of concepts in order to understand proving. In this case, he is probably justified in avoiding such theorizing, since it is as he said, viz., "concept" is vague.

9. Let us follow Wittgenstein's suggestion in the opening quotation of this Chapter. Let us call necessary truths such as those discussed above: essential truths or statements of essence. The singular ones will state essential features of individuals and the general ones, such as B AFTER IT IS PROVED, state essential features of kinds of things. I suggest that we regard them as statements of essence for the following reasons. First, I think that if any statements could be called statements of essence, these can. The singular ones tell us of a feature an individual must have to be of the kind that it is. Since, if an individual were not of the kind that it is, it would not be the individual that it is; these singular essential truths tell us of a feature that an individual must have to be what it is. The general ones tell us of a feature that any member of a kind must have to be of that kind. Secondly, by regarding them as statements of essence, we can better understand remarks such as "The mathematian creates essence" from I-32. It will also help us understand two passages from his *Philosophical Investigations*. These passages are: "*Essence* is expressed by grammar" from *PI*-371 and "Grammar tells us what kind of object anything is. (Theology as grammar.)" After the

proof of B, there is an implicatory link between the concepts "being a Bismarck" and "having 8 letters", this link can be called a grammatical link, and knowledge of this link tells us in part what a "Bismarck" is. So it is not eccentric to say that a proof of B results in a new bit of grammar and that knowledge of this bit of grammar is knowledge of what a "Bismarck" is. But I am not suggesting that one could fix the grammar of "God" in the same way as we can fix the grammar of "Bismarck-configuration".

10. Let us also convince ourselves of, or at least try to appreciate, Wittgenstein's claim in the opening quotation that B as A PROVED PROPOSITION is synthetic *a priori*. As it stands B is not analytic. It does not have the form of a logical truth any more than "The number of Tom's children is 8" has the form of a logical truth. Of course, a person who is convinced that the only necessity is logical necessity will not be impressed by the fact that, as it stands, B does not have the form of a logical truth. Such an opponent of the synthetic *a priori* may claim that B can never be proved, i.e., B can never be more than an extremely well justified contingent empirical truth about words. Or else, he may argue that when properly analyzed B would be rewritten in a form which is the form of a logical truth. In face of the inconceivability of the denial of B, it is sheer dogmaticism to claim that B cannot be proved. (Recall that I am using "proved" to mean established as necessarily true.) So our opponent of the synthetic *a priori* could honestly argue only that when properly analyzed B would become analytic. In the proposed proper analysis "Bismarck-configuration" would be defined so that B would be transformed into an analytic sentence which may read like the following. "The number of letters in a word which has amongst other features 8 letters is 8." Such an analysis can be given but it will not show that B is really analytic. For one: there is no definition of "Bismarck-configuration." So one shall have to be made up. In order to make up such a definition of "Bismarck-configuration", we have to recognize essential features of "Bismarck" in order to know what to put into the definition. Recognition of essential features requires recognition of the very kind of neces-

sary truths which the definition is supposed to show are eliminable. For instance, how would one know that a definition of "Bismarck-configuration" should be such that it states or implies that the number of letters in any token of "Bismarck" is 8? I believe that a proof such as that which we developed from Picture 1 would be necessary to determine that a definition of "Bismarck-configuration" must state or imply that "Bismarck" has 8 letters. Even if one accepted the conventions for spelling "Bismarck" as a definition of "Bismarck-configuration", one would still not be able to avoid appeal to essential truths in order to transform B into an analytic sentence. The conventional rules for spelling "Bismarck" do not contain an explicit provision that we write down 8 letters. To know that these conventions do imply that "Bismarck" has 8 letters, we would have to follow these conventions to construct a token of "Bismarck", and then do as we did in Picture 1, viz., read-off B as an essential truth. In general, any configuration of letters or simple structure can have the term standing for it defined only after we have determined what is essential to the structure. So use of definitions will not enable one to show that sentences like B are really analytic.

This brings us to another reason for saying that B is not really analytic. B can also be used to state an experimentally justified claim about the number of letters in any token of "Bismarck". If someone did not try to think of B being false, he would regard B as nothing more than an experimentally justified truth about a word. If B were really analytic, one would have to hold that B is never really suitable for being used to make experimental statements. But it can be used to state the result of an experiment; thus B is not really or implicitly analytic. A third reason for saying that B, when looked at as necessarily true, is synthetic *a priori* is that our way of establishing its necessity seems very much like the way Kant says that we would establish a synthetic *a priori* judgment. (In Section 2 of his *Prolegomena to Any Future Metaphysics,* Kant claimed that to establish $7+5=12$ as synthetic *a priori* we have to inspect empirical situations for applying "five added to seven".) I think that my arguments about definition show that an analysis of the concept of being a "Bismarck" will not show us that B must be true. However, if we represent to visible intuition, i.e., sight, a situation for applying "Bismarck-configuration" and "having 8 letters" we can judge that these two concepts are connected because of what our sensuous intuition, – again sight –, shows us in Picture 1.

I am not, of course, claiming that a strict finitist will also hold the Kantian positions that space and time are forms of sensuous intuition. The Kantian positions on space and time are philosophical theories to account for and justify what is done in a strict finitistic proof. A strict finitist can avoid these positions by either advancing different philosophical theories or by trying to philosophize without the development of any philosophical theory. In other words, I am not suggesting that the Kantian and the strict-finitist would give the same answer to: How is pure mathematics possible? I am only suggesting that they would agree on how a kind of judgment, called synthetic *a priori* by Kant, is established. But agreement on this last point vindicates calling propositions proved by a strict finitists "synthetic *a priori*". So, I hope that I have shown the appropriateness of Wittgenstein's suggestion that after a proof B is synthetic *a priori*.

11. Let us now turn to an analysis of what we did when we gave a strict finitist proof of B. Or as we have just seen, we could say that we want to analyze what we did when we established that having 8 letters was of the essence of "Bismarck" as well as saying that we want to analyze what we did when we made the synthetic *a priori* judgment that B. We shall go through the analysis by analyzing what we do when we give a strict finitistic proof of "The reverse of OVER is REVO" rather than analyzing our proof of B. In this way we get a better idea of the range of sentences in a strict finitist's paradigm kind of sentence. This is helpful since there is no precise characterization of what kind of subject matter strict finitistic paradigm sentences are about. Although I suspect that all, or most, strict finitistic sentences have, or can be given, the form: The X of Y is Z. Also the necessity of "The reverse of OVER is REVO" is even more obvious than that of B. So its proof is even more suitable for an analysis of how we establish necessity.

12. The opening sentence of the preceding paragraph shows that I am using "strict finitist" ambiguously. In some cases, I use "strict finitism" to stand for a philosophy of mathematics. In other cases, I use it to stand for a way of doing mathematics. The way of doing mathematics that I call "Strict finitistic" is that portion of mathematics of which a strict finitist as a philosopher can give an account. I believe that throughout this essay it will be clear from the context whether I am talking about a kind of mathematics or a philosophy of mathematics.

13. Let us now turn to an analysis of a proof of "The reverse of OVER is REVO". (I shall abbreviate this sentence as RO.) The proof pattern of RO is in Picture 2 below.

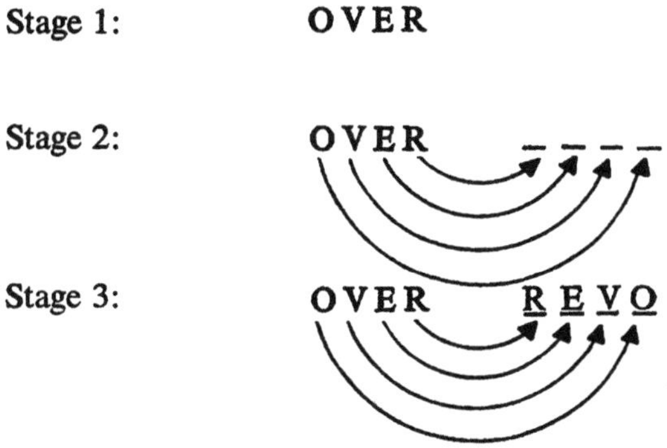

14. The first stage in a strict finitistic proof of a proposition of the form: "The X of Y is Z" is to construct, or represent to our senses or imagination, a typical instance of the operand or Y term. So at Stage 1 of Picture 2 we construct a typical OVER. But not just any OVER will do.

The typical instance of the operand, – in our case the typical token of OVER –, must be perspicuous. To be perspicuous a figure must meet certain requirements of size, simplicity, and state of motion. If a figure is perspicuous, it is not so large that we must use inductive reasoning to identify it as being an instance of the operand term; nor is it so large that we would have to use inductive reasoning to determine that we have actually applied to it the operation which we intended to apply to it. For instance, if an OVER had each letter as large as a sky-scraper, it would not be perspicuous unless we viewed it from a great distance. (But if we viewed it from a great distance, we would have a reasonably small OVER before us.) If we were to stand close to one of the sky-scraper sized letters, we could not see all the letters together as a group; thus we would have to use inductive reasoning to determine that this huge letter is in an OVER. Certainly, with letters of this size, we could use only inductive reasoning to determine whether a rearrangement of them was a reversal of them. Similarly, a perspicuous figure is not so small that we need to use inductive reasoning to know what kind of figure it is or to know whether some manipulation of it is the manipulation we intended to perform or have performed. Closely connected with the requirements on its size is the requirement that a perspicuous figure not be so complex that we cannot, without induction, identify it or the operations we apply to it. This last requirement is connected with size because often an enlargement will produce the required simplicity. (Of course, the enlarged figure is a new figure for the first stage of the proof; it is not the complicated small one.) What I mean by saying that a certain state of motion is required for perspicuity is that the figure must not be undergoing physical changes so that we need to use inductive reasoning to identify it or operations we apply to it. For instance, if an OVER were whirling around in a circle, we could not directly, – with a single look –, identify it as an OVER.

It may help to cite some sentences in which the operand is not perspicuous for being of a kind and for the operation talked about. Wittgenstein's II-16 suggests two examples, although he does not use these examples as I use them here. The examples are: "The number of peas in this vessel is the number of his shillings" and "The number of letters in the Song of Songs is the number of his shillings." We cannot tell with absolute certainty that we have counted all the peas in a vessel. We cannot tell without induction that any given printing of the Song of Songs has every letter in it that should be in it. Again, we can never be absolutely certain that we have counted every letter when we try to count the letters in a copy of the Song of Songs. So the peas in a vessel are not perspicuous for the operation of counting them, unless there are only a few. And also, no copy of the Song of Songs is perspicuous for being of the kind: copy of the Song of Songs. (Wittgenstein cites these examples to remind us that although referring to numbers in these ways is worthless for mathematical purposes, i.e., giving proofs, they are legitimate expressions of empirical data.) A lump of ground beef provides another example of an object that is not perspicuous for being one of a kind and for a certain operation. Consider the sentence: "The number of golf-ball size meatballs formable from a pound of hamburger is 8." Certainly, one cannot tell without induction that a lump of ground-beaf is a pound of hamburger. Nor can one be absolutely certain that when he forms meatballs he is forming golf-ball size ones. Another value of citing these examples is that it shows us that not all sentences can be made into necessary truths by strict finitistic proofs.

I should note that my main textual clue in developing this concept of "perspicuity", as well as those of "memorability", and "surveyability" came from the last paragraph of II-39 where he wrote:

> it must not be necessary to make a physical investigation of the proof-configuration in order to shew us what has been proved.

I also followed a similar clue from III-41.

Why does a strict finitist need perspicuity? A strict finitist, – as a mathematician –, wants to establish a claim of the form: "The X of Y MUST be Z." And he wants to establish this by reading it off from a picture showing that the X of a particular Y is a Z. If he is not certain that he has a Y in his picture he, of course, will not see from that picture what the X of Y must be. For instance, looking at something you are not sure is an OVER will not give good evidence what the reverse of an OVER IS, let alone show you what it MUST be. Similarly, if you could doubt that a procedure you pictured as taking the X of Y really was taking the X of Y, the picture would not convince you of much about the X of Ys.

To me it seems appropriate to compare perspicuous figures

with sense data. They are like sense data insofar as their properties which interest us are concerned. Their seeming to have them is sufficient for their having them. And their seeming not to have them is sufficient for saying that they do not have them. Thus a very large OVER does not seem at a glance to be an OVER, so as a perspicuous figure it is not one. Also if a figure seemed, to all who looked at it, to be an OVER, it would suffice as an OVER for a proof. However, perspicuous figures are not like sense data insofar as they are not objects posited to solve epistemological problems.

15. Wittgenstein's criterion for perspicuity was ease of reproduction. In II-1, regarding his claim that a mathematical proof must be perspicuous, he commented:

> 'A mathematical proof must be perspicuous.' Only a structure whose reproduction is an easy task is called a "proof." It must be possible to decide with certainty whether we really have the same proof twice over, or not. The proof must be a configuration whose exact reproduction can be certain.

If we can be certain that a figure F-2 is a reproduction of another figure F-1, we must be certain about what F-1 was and certain about what F-2 is. If we are certain about what the figures are, we have found this out without induction. Now, knowing what these figures are cannot, in this context, be knowing every feature of these figures or even very many of their properties. For instance, we cannot tell without induction the exact color of an O, its exact size, or its curvature although we can be certain that it is an O. Wittgenstein realized this because he continued II-1:

> Or again: we must be sure we can exactly reproduce what is essential to the proof. It may for example be written in two different handwritings or colours. What goes to make the reproduction of a proof is not anything like an exact reproduction of a shade of colour or a hand-writing.

For instance, both a red OVER and a blue OVER are reproductions of what we have at Stage 1 of Picture 2, as long as we can be certain that we still have OVERs and can be certain that the operation we call their reverse really is their reverse, because these are the features essential to a proof of RO. So I submit that knowing what figures are, when we are talking about a proof, is knowing that they are figures of a certain kind and being able to identify with certainty that an operation we do apply to them is the operation referred to in the sentence to be proved. Hence, I conclude that Wittgentsein's criterion for perspicuous figure is the same as mine.

In V-17 there is a discussion of perspicuous figures which is very much like mine. But I must admit that in V-17 as well as in II-1 he was talking about the entire proof pattern being perspicuous instead of applying "perspicuous" only to the paradigm of the operand. However, in section 20 we will see that this difference with Wittgenstein is not serious.

16. The second stage of a proof of RO is experimental. We experiment to find out what does happen if we put the last letter of OVER in first place, the second to the last in second place, etc. (In Picture 2, I picture the experimental stage as Stage 2. In actual practice Stage 2 and Stage 3 would be indiscernible since when you experimented as to what results when you put last first etc., you would actually write down the letters in reverse order. Still there is a difference between Stage 2 and Stage 3. The difference is that Stage 2 pictures getting the result experimentally while Stage 3 pictures describing the experimentally obtained result.) Stage 2 is experimental in the sense that the immediate consequence of this stage is only that we see that a certain result does follow rearranging the letters. We could stay at this empirical stage and then, after we described our experimentally obtained result, settle for RO as a mere empirical truth.

17. At this stage it is appropriate to discuss the concept of memorability, since a proof pattern gets the feature of memorability from the kind of operation performed at the experimental stage. It is the picture that we draw of our experimenting that must be memorable. A picture we draw of our experimenting, on what happens if we apply a certain operation, is memorable if we can tell without induction that what we have pictured is the

operation we intended to picture. For instance, Stages 2 and 3 of Picture 2 are memorable pictures of reversing letters because once we write the lines down to represent reversal there can be no doubt that they show reversal. But 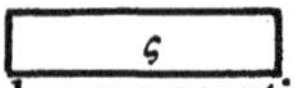 is not a memorable picture of what we get if we apply an operation such as: The result of drawing a ς through the center of  is ________. (This rectangle example comes from I-70.) We cannot be certain that the wavy line we draw through the rectangle is the same kind of wavy line as the one we intended to draw in order to apply the operation: The result of drawing ς etc. We will not be certain because we cannot remember exactly what the wavy line looks like. So if we do this experiment again, viz., draw ς through a rectangle, we need induction to know whether or not we have really performed the same experiment again. Hence, here, as in the case of perspicuity, to say that we do not need induction to tell whether two pictures are the same is to take ease of reproduction as a criterion of memorability. This example with the wavy line also shows that it is not odd to call this feature of proof patterns "memorability".

A strict finitist, as a mathematician, needs memorability because if he cannot be certain that he ever pictures the X of , no picture which is only possibly a picture of the X of a Y will show him what the X of Ys MUST be.

18. At the third stage we establish two empirical truths. Because of what we see we judge that the reversal of the OVER in Picture 2 is REVO. Then, since we have taken the OVER there as a typical OVER, we also make the general statement, on the basis of this experiment, that the reversal of OVER is REVO. It must be emphasized, though, that RO is not used here to express a necessary truth; at this stage it is used only to express an experimentally justified belief. I represent this third stage of establishing the truth of "The reversal of this OVER is REVO" as filling in the blanks from Stage 2; I was trying to picture seeing that as a matter of fact REVO came out. I did not picture generalizing the experimental result. Still, notice that there is a difference between the singular statement describing the result of the experiment and RO as an empirical generalization. The sin-

gular statement is based on the sensory evidence in Picture 2 while RO, as an empirical generalization, is inductively grounded in the singular statement plus statements that the OVER in Picture 2 is a typical OVER. The point is that in the empirical realm general statements are based on singular ones. We will see in section 23 that there is not such a relationship between the mathematically general and the mathematically singular. However, our main interest is not in the empirical statements, – the IS statements –, which are justified by the experiments that we picture. I call attention to the empirical statements which are justified by the experiments pictured in strict finitistic proofs so that I can show that strict finitists make a transition from IS to MUST. Study of this transition may provide a clue on how to make the transition from IS to OUGHT. (For a discussion of IS to MUST and IS to OUGHT transitions see my: "Deduction and Intuitive Induction" in the March 1966 *Journal of Philosophy and Phenomenological Research.*) In this essay, however, our main interest in the third stage is in the picture which is drawn there. This picture is the completed proof pattern and it must be surveyable.

19. A picture of an operation being performed is surveyable if it shows us the result being part of applying the operation or getting the result being the same as applying the operation. (There is no significant difference between part of applying an operation and being applying the operation; it all depends upon the description of the operation. For instance, having an R before an E is only part of applying the reversal of OVER while having REVO is applying the reversal of OVER. But we could easily describe an operation on OVER such that having R before E would be applying the operation. So in the future I shall talk only of having the result being the same as applying an operation.) I believe that I am using "surveyable" as Wittgenstein used it. In II-55 he wrote:

"Proof must be surveyable": this aims at drawing our attention to the difference between the concepts of 'repeating a proof' and 'repeating an experiment'. To repeat a proof means, not to reproduce the conditions under which a particular result was once obtained, but to repeat every step *and the result.*

Is it not true that if someone drew another picture of reversing the letters of an OVER he would draw what is in Stage 3 of Picture 2? And Stage 3 of Picture 2 contains the result of the reversal as well as a picture of the reversing. So isn't Stage 3 of Picture 2 surveyable in Wittgenstein's sense? Doesn't it also show having REVO as identical with reversing the letters? So isn't it surveyable in my sense?

Of course a strict finitist wants surveyability because he wants to see what must result from applying a certain operation. If he can see that a certain result is one and the same as applying the operation, he will see that the operation must have that result, viz., he will have a proof. I think that we can now see why Wittgenstein, if he was a strict finitist, still held in *RFM* III-36 as he did in 6.2 of his *Tractatus* that mathematical statements are a kind of statement of identity. They tell us that certain processes are the same as certain results. A strict finitist also wants surveyability because as Wittgenstein put it in II-22: "Proof, one might say, does not merely show *that* it is like this, but: *how* it is like this. It shows *how* 13 + 14 yield 27." Why does a strict finitist want a proof to show him how an operation gives a result? Assume that one does not know how an operation gives a result. Then whatever happens after he starts applying the operation is a possible result. The consequence most frequently following may be what one picks as the result of applying the operation; call it Z. But he will have no criterion for saying that the operation is such that it gives this result. As long as one cannot definitely identify what "The X of Y" denotes, he does not really understand the phrase. If you do not know how taking the X of Y gives Z, you cannot definitely identify Z as the result of taking the X of Y, i.e., you cannot definitely identify the X of Y. Now knowing how an operation gives a result Z is seeing that getting Z by taking the X of Y is what the X of Y is. (Isn't it true that Stage 3 of Picture 2, by showing HOW reversing the letters of an OVER gives a REVO, shows that the reversal of the letters of an OVER IS a REVO?) Hence, a strict finitist wants to know how an operation gives a result so that he can know what the operation is. In effect, for the strict finitist the proof defines the operation or gives a model or paradigm for what it is to apply the operation. He would agree with Wittgenstein's way of putting it in II-24:

> The proof is our model for a particular *result's being yielded*, which serves as an object of comparison (yardstick) for real changes.

And surveyability is necessary to see how an operation gives a result and hence necessary for a particular picture to be a model of what it is like to apply an operation.

In light of this discussion of surveyability, we can understand a remark made by Wittgenstein in both his *Tractatus*, at 6.1261, and in *RFM*, at I-82. The remark is that in mathematics process and result are equivalent. This means that knowing what a mathematical operation – the X of Y – is knowing the result of taking the X of Y. So this remark expresses the position that we have a mathematical operation only when we can identify it with its result. At least a sufficient condition for identifying an operation with its result is that we have a paradigm of how that operation gives its result; and this we have if we have a surveyable picture of an application of the operation.

20. Before I begin a discussion of the next two stages, which turn a strict finitistic proof pattern into a proof, it may be helpful if I discuss the relationships between the concepts of perspicuity, memorability, surveyability, and then compare my use of them with Wittgenstein's.

If we have a picture of an operation, – the X of Y –, being applied to a figure of kind Y we have the following results. (a) If the figure is perspicuous as a Y and perspicuous for applications of taking the X of Y, then the picture is a memorable picture of taking the X of Y. We have (a) because if a particular figure is perspicuous for application of an operation we can tell without induction that we have applied the operation to it; and this is the criterion for memorability. (b) If the picture is a memorable picture of an application of the operation, – the X of Y –, the figure in the picture to which it is applied is perspicuous as a Y and perspicuous for applications of this operation. The argument for

(b) is obvious. If we can tell without induction that we have a picture of applying the X of Y, we must be able to tell without induction that we have a Y and that we have taken the X of it. (c) If the picture of the operation is surveyable, then the figure to which it is applied is perspicuous and the picture of the operation is memorable. The justification of (c) is also brief. If the picture shows us that the X of Y is identical with Z we must be certain that, i.e., know without induction, that we have a Y and have pictured taking the X of it. We cannot identify one thing, – Z –, with something which we cannot definitely identify. (Incidently, if the picture is surveyable, the Z pictured will be perspicuous as a Z.) (d) If the figure in the picture is perspicuous as a Y and perspicuous for applications of the X of Y, the picture of taking the X of this Y is memorable and, if the Z pictured is perspicuous as a Z, the picture is surveyable. The argument for (d) is a bit more complex. If we are absolutely certain that we have a Y, a X and absolutely certain that our picture of getting Z as a result of taking the X of the pictured Y is a picture of taking the X of that Y, then we can see taking the X of that Y as being identical with having Z. If we do not see getting Z as identical with applying the operation, we admit that Z did not have to be produced by application of the operation. But the admission that applying the operation did not have to produce Z is inconsistent with the assumption of memorability. I.e., it is inconsistent with the assumption that we were absolutely certain that the picture of getting a Z by taking the X of the pictured Y was a picture of taking the X of that Y. There is an inconsistency because if we are not certain that a Z had to be produced when we took the X of the pictured Y, the presence of a Z is only some evidence that we took the X of the Y.

A consequence of (a) through (d) is that perspicuity, memorability, and surveyability are equivalent concepts when applied to a picture which shows a Z being produced by taking the X of a Y, if the pictured Z is perspicuous as a Z. In strict finitistic proofs, the Z, – the pictured result –, will always be perspicuous as a Z. So whenever there is occasion to use these concepts they are equivalent concepts. This is why in *RFM* we will find no sharp distinction between these concepts. Nonetheless, I believe that a detailed investigation of these concepts helps in understanding what Wittgenstein was trying to say with them.

21. Let us now analyze how we go from the perspicious, memorable, and surveyable picture of Stage 3 to obtain necessary truths. Recall that at Stage 3 the only truths that we have are: The singular "The reversal of the letters of the OVER pictured in Picture 2 is the REVO pictured there" and RO as an empirical generalization about OVERs. Of course we also possess a surveyable picture of our grounds for the above two empirical, or is-statements. It remains, though, to transform these is-statements into must-statements. The transformations will take two more stages.

22. On the fourth stage of a strict finitistic proof the prover makes the move from the singular "is" to the singular "must". The fourth stage of our example is the movement from "The reversal of the letters of the OVER pictured in Picture 2 IS the REVO pictured there" to "The reversal of the letters of the OVER pictured in Picture 2 MUST be a REVO." The prover makes the move from the singular "is" to the singular "must" because he can survey Stage 3 of Picture 2 and see that getting REVO is what it is to reverse the letters of that OVER. As Wittgenstein put it in II-39: "This *must* come out – then this is not for reasons which we do not see." Of course, what we see when we see what MUST be is what we see after appreciating the surveyability of Stage 3.

I grant that the product of the fourth stage is so uninteresting for non-philosophical purposes that its existence is unknown. However, calling attention to the fourth stage is philosophically interesting because it shows the vital role of diagrams in proofs. The diagrams, proof patterns are not rough sketches of something else; in the proof we determine what must be true of the diagrams themselves. A strict finitist treats the diagram as a platonic form, if you will, of the kind of figure talked about. What is of the essence of it holds true for all members of the kind. But before we can determine what must hold true for all members of the kind we must determine what holds true for the form of the kind, and the form of the kind is a typical member of the kind in the way OVER is a typical OVER.

It is important to note that Stage 4 as well as Stage 5 are parts of a strict finitistic proof although they cannot be written down in any way. They are, so to speak, within the person doing the proving. For a strict finitist a proof is a production of a necessary truth, and this production is a resultant of a proof-pattern and a prover. This is what Wittgenstein meant in II-42 when he wrote: "It is not something behind the proof, but the proof that proves." The proof is not some description of evidence such that if we looked at the evidence behind the statements in the proof we would see that the conclusion is warranted. No, the proof is the activity we perform in the presence of the proof-pattern. Or to say it in another way: proving is a response to that bit of reality which is the proof-pattern. A strict finitist responds to the proof-pattern by identifying a certain result with an operation. One could even say that his response is the production of a new standard for measuring whether or not a particular operation has been applied. As Wittgenstein wrote in II-27:

> I am trying to say something like this: even if the proved mathematical proposition seems to point to a reality outside itself, still it is only the expression of acceptance of a new measure (of reality).

The fact that a strict finitistic proof requires conditions on the part of the prover seems unfortunate since it leaves a strict finitist open to the question Wittgenstein asked himself in II-42: "Are you really saying anything but: a proof is taken as *proof*?" Clearly, for a strict finitist, being taken for a proof, – here "proof" refers to proof-pattern –, is necessary for there to be a proof. For instance, a prover must be convinced by Stage 3 that REVO must come out for Stage 3 to be a proof-pattern for RO. It sounds even more unfortunate, but for a strict finitist being taken for a proof is sufficient for being a proof. A strict finitist would concur with Wittgenstein's remark in II-39:

> What convinces us – that is the proof: a configuration that does not convince us is not the proof, even when it can be shewn to exemplify the proved proposition.

If proving is producing a necessary truth in the face of a surveyable pattern, being convinced, i.e., appreciating the surveyability of the pattern, produces the necessary truth. It sounds bad to say that being convinced is necessary and sufficient for proof because this seems to imply that any proposition can be proved, i.e., legitimately proclaimed as a necessary truth, and that there can be no mathematical mistakes in the sense that there cannot be an incorrect proof. The restricted supply of surveyable pictures prevents the first consequence. For instance, one simply cannot get a pattern to prove: The number of golf-ball size meatballs formable from a pound of hamburger is 8. In Chapter III, section 24 we will see how a strict finitist accepts the consequence that there can be no incorrect mathematical proof.

23. In a proof, a strict finitist engages in an activity of producing necessary truths. However, we have not yet analyzed this production. We shall do so now when we discuss the fifth and final stage of a strict finitistic proof, viz., the generalization stage.

What entitles us to go from the singular to the general? In this case, we clearly cannot do as we do in the empirical realm; we cannot use the singular statement as inductive grounds for the general statement. We cannot do this because inductive grounds are, of course, irrelevant for necessary truths. It is the perspicuity of the instance of the operand which permits us to generalize. So since perspicuity is a feature of a singular thing, – the instance of the operand –, general necessary truths are also grounded in truths about individuals. Let us see how they are grounded by considering Picture 2. The OVER in Picture 2 is perspicuous, i.e., we can be absolutely certain that it is an OVER. Hence, we are absolutely certain that there is before us what we would identify whenever we identified an OVER. Consequently, when we see this single perspicuous OVER we see something general or universal, because what we would identify when we identify any OVER is general or universal in that it must be common to all OVERs. Wittgenstein's reproducibility criterion for perspicuity shows that he also believed, – implicitly –, that what we see in a perspicuous figure is universal because what we reproduce is more instances of an universal. Nonetheless, even if what is perspicuous in a figure is a universal, it is in the figure. (Or to try to avoid the metaphorical "in", we could say that the universal is the perspicuous figure from a point of view in which it is looked at solely as a member of a kind Y. In our example, the point of view from which the OVER in Picture 2 is an universal is that point of view in which we look at it as an OVER and nothing else.) So if the universal is in the singular, or is thought about by thinking abstractly about the singular, we must talk about the singular object. That is the function of Stage 4. At Stage 4 we establish a necessary truth about the universal as it is in the singular. But since the universal, although in the singular, is still an universal what we establish MUST be true of it somewhere is true of it anywhere. So the generalization from "The reversal of the OVER in Picture 2 must be a REVO" to RO with a "must" in it, i.e., "The reverse of OVER must be REVO" is grounded in the proof-pattern because the OVER which we see there is any OVER. Really there is no change in subject matter when we switch from the singular to the general statement. In both statements we are talking about the same universal; only in the general statement we ignore the fact that the universal is in a single object. In effect there is only a verbal difference between the singular and the general in the mathematical language of a strict finitist, but the general statements are the ones he uses.

I must emphasize that the preceding speculation about perspicuity as seeing an universal in the particular is not Wittgenstein's doctrine. I think that it is compatible with his positions, if not with his approach to philosophy. But to my knowledge, he did not explicitly hold it.

24. Now the question is: For what are the general necessary truths used? I believe that Wittgenstein gave the answer when he wrote in II-30:

> I go through the proof and say: "Yes, this is how it *has* to be; I must fix the use of my language in *this* way.

in III-29 when he wrote:

> What is the transition that I make from "It will be like this" to "it

must be like this"? I form a different concept. One involving something that was not there before.

and in II-26 when he wrote:

> Let us remember that in mathematics we are convinced of *grammatical* propositions; so the expression, the result, of our being convinced is that we *accept a rule*.

Let us interpret these quotations by reference to our proof about the reverse of OVER. As a result of this proof, we set up an implicatory connection between the concepts "being the reversal of the letters of OVER" and "being a REVO". If one does not like talk about concepts he could say that the person who proves sets up an identity between the two terms "the reversal of the letters of OVER" and "a REVO". This conceptual connection or identity of terms is not put into language because the referents of these concepts or terms are connected as they are said to be connected in the proved proposition. This is not a denial that OVERs and REVOs are so connected that if an OVER is reversed a REVO is always produced. It is denied only that the proved proposition describes some previously existent reality for whose existence the proof supplies (conclusive) evidence. This denial is a claim that the change from "is" to "must" is not a change from one kind of description to another. For instance, we do not switch from describing contingent connections to describing necessary connections. The is-proposition adequately describes the connections which exist outside or prior to the proof, and the experiment which the proof-pattern pictures provides convincing, if not inductively adequate, evidence for the existence of the connections. The must-proposition expresses the prover's determination on how he will talk about the connections in the future. He resolves to say in the future that REVO is the same as the reverse of OVER; he forms this resolution because of the surveyability of the proof-pattern.

If my use of "is-proposition" and "must-proposition" is confusing, the confusion can be removed by considering RO. When I talk about RO as an is-proposition, I am talking about it as an empirical statement describing the result of an experiment. When

I talk about it as a must-proposition, I am talking about it as a proved proposition. To have a must-proposition it is not necessary to literally change the "is" to "must be". The "must" is usually implicit.

Of course, one could interpret to move from "is" to "must" as a transition from one kind of description to another. He could suggest that in the first case we describe a contingent connection while in the second case we describe a necessary connection which we have seen in the surveyable proof-pattern. To interpret must-statements as descriptions of necessary connections is not incompatible with admitting that as a result of them we create an implicatory connection between concepts or terms. The difficulty in interpreting must-statements as descriptions of necessary connections is that one then has to develop an epistemology with a place for intuition as a faculty for recognizing these necessary connections, viz., truth conditions for the must-propositions.

25. Before we turn to a strict finitist's answers to the three groups of questions on the philosophic foundations of mathematics, I shall speculate on how action by one person, – the prover –, can produce a rule in language. Such speculation may help in interpreting the three passages I just quoted from Wittgenstein.

Of course, one man cannot produce a conceptual connection or an identification of terms in language. A language is not a man's private possession with which he can tamper at will. What actually happens as a result of a proof is that many people grant that what does come out must come out. As a result of many people saying, or just thinking that it must come out, e.g., that the reverse of OVER is REVO, they refuse to call anything the reversal of OVER unless it is a REVO. And if asked they will call REVO the reversal of OVER. It is this agreement on what they appreciate in proof-patterns and the corresponding agreement in usage which sets up rules of implication and conceptual connections in language. In other words, it takes more than one to lay down a track in language, but still there are tracks in language because several individuals each try to lay down the same track.

In II-65 to II-76, I submit that Wittgenstein made the same point about the need for agreement. But in II-65 to II-76 he was primarily concerned with arguing that although we need the anthropological phenomena of agreement in appreciating necessity, agreement in use, and, as he added in V-6, agreement in an intolerance towards people differing from our use, mathematical propositions do not assert the presence of these agreements.

Of course, in sophisticated cultures there are specialists on appreciating what must be, viz., mathematicians and logicians, and amongst the specialists there are experts and geniuses. In such a sophisticated culture, it would not be too inaccurate to talk about a small group altering a part of the language for everyone else. Also since in the sophisticated culture people are aware of the effect of appreciating what must be, i.e., a rule is developed, they may well not wait for the rule to evolve from agreement in use. They may explicitly make the rule when the theorem is proved by one of the few specialists. But in simple cultures the appreciation of necessity, – proving –, and the alteration of language is the achievement of the entire culture. For instance our elementary arithmetic is not the product of specialists; millions of laymen fixed language so that a group of five can be referred to with either "five" or "two and three". Really in a simple culture the existing mathematics is more rational for the majority of the people than in a sophisticated culture where the majority of the people accept explicit mathematical rules from the specialists. In a simple culture the people through their use of language develop the mathematical rules of their language and their usage is based on what they think must be the case.

The speculation, just advanced, on how individual's views on what must be result in a change in language is not incompatible with the following remark in III-8.

> We judge identity and agreement by the results of our calculating; that is why we cannot use agreement to explain calculating.

Calculating is proving, viz., the determination of what must be by an individual prover. There is no need to talk about agreement with anyone when we talk about the determination of necessity.

It is the proof that proves; not something behind the proof such as agreement with others that makes the proof work. However, when we come to discuss the usual effects of the individual's proofs, we are talking about a natural phenomenon, viz., alteration of language. So in the preceding paragraph I was doing a bit of casual anthropology rather than philosophy. This anthropology is compatible with the just quoted remark from III-8. It is also consistent with Wittgenstein's objections to explanations in philosophy which we noted in Chapter I. I was only trying to explain the natural phenomenon of alteration of language by what Wittgenstein in I-35 called a great and interesting agreement in calculation. Further values in even sketching such an explanation are as follows. Pointing out the need for agreement in calculation reminds us that language is not something an individual can make up for himself. Pointing out that there is agreement in calculation reminds us that worry about each individual trying to develop his own mathematics, – and thus having none –, is baseless.

26. Let us now leave anthropological speculation and return to philosophy. How would a strict finitist answer the questions on the philosophic foundations of mathematics? A strict finitist would answer (Ii) by replying that statements made in pure mathematics are not about any objects. So he would not have to worry about (Iii) and Iiii); hence he has avoided platonism. I should emphasize at the outset of this discussion of strict finitism and mathematical objects that their denial of mathematical objects is not inconsistent with the discussion of universals in section 23. One can consistently say that in a proof we pay attention to universals without saying that we discover truths about connections between universals and without saying that the order and connection of universals provide truth conditions for our proved propositions. To say that we pay attention to universals is to admit that we talk about kinds of figures, but saying this does not imply that we discover what must be true of a kind rather than invent what must be true of a kind. But let us examine a bit more closely what entitles a strict finitist to answer that mathematical propositions, which are proved by strict finitistic means, are not about any objects, viz., do not have truth conditions. To

do this let us look at a strict finitistic proof of: The sum of two and three is five.

The third stage of a strict finitistic proof of $2 + 3 = 5$ could be:

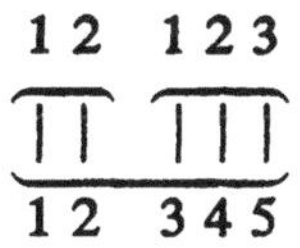

Let us call this proof-pattern Picture 3; it is a slight modification of a pattern which Wittgenstein presented in I-62. In Picture 3 we have perspicuous figures for applying "two", "three", "five", and a memorable picture of counting two and three together. The picture enables us to survey that the counting together of two and three is counting five. But there is nothing perspicuous and memorable, and nothing is surveyed except Picture 3. So the only objects that there are are the individuals and their properties in Picture 3. The singular is-statements tell us how it is with them. Or, we could say that the only truth conditions that are before us are those conditions which are truth conditions for singular statements such as: The counting together of the two marks and the three marks in Picture 3 gives a count of five marks. There is no need to talk about a number two and a number three standing in some necessary relation with the number five. All that is necessary for the proof is that we have perspicuous figures to which we can apply the adjectives "two", "three", and "five". If I sound dogmatic in claiming that we do not need to look at numbers to prove that $2 + 3 = 5$, one must ask himself what he has to look at besides a picture such as Picture 3. Then he must ask himself what he sees in such pictures that can be called a number two rather than a group with two members. One could look at it this way. A strict finitist denies that there are mathematical objects because he can establish mathematical theorems by using mathematical terms as adjectives applicable to observable figures, but he does not need to use them as nouns allegedly standing for some objects.

A comparison of strict finitistic theorems with rules of measurement and a comparison of strict finitistic views of proof with

Kant's and Berkeley's views on establishing necessity may clarify strict finitism's position on mathematical objects.

27. In V-15 Wittgenstein claimed:

'There are 60 seconds to a minute.' this proposition is very *like* a mathematical one.

This proposition about time is like $2 + 3 = 5$ in that both are rules on identifying terms and both require some empirical conditions for being made. For instance, if nature was so chaotic that men could not observe small regularities, they could not talk about seconds. Similarly, if men could not visualize a group of more than four, they would not be able to establish that $2 + 3 = 5$, and they probably would have no use for it if they introduced it by convention. So their use and conditions for their use are the same. They differ primarily in their origin, viz., source of their introduction (gradual) into language. "There are 60 seconds to a minute" is introduced by convention while $2 + 3 = 5$ is proved. We have no inclination to say that "There are 60 seconds to a minute" is true because it corresponds with its truth conditions and its truth justifies the convention. For our purpose, the question is whether or not the fact that $2 + 3 = 5$ is proved differentiates $2 + 3 = 5$ from "There are 60 seconds to a minute" enough so that we should say that $2 + 3 = 5$ is about objects. (I should be happy to frame the question as: Although analytic truths do not have truth conditions, do synthetic *a priori* truths have truth conditions?) A strict finitist would answer that the fact that we can present surveyable pictures of $2 + 3 = 5$ but not for "There are 60 seconds to a minute" accounts for the fact that one gets put into our languages as a result of a proof while the other requires explicit convention. At least we cannot get a surveyable picture of 60 seconds being a minute before the convention is made. I could also have said that $2 + 3 = 5$ becomes a rule of language by a convention which evolves because most people have a proof of it while "There are 60 seconds to a minute" requires an explicit convention to become a rule of language and this convention is not the effect of a proof.

28. In Section 11 of Part One of his *Prolegomena to Any*

Future Metaphysics, in L. W. Beck's translation in the Liberal Arts Press, Kant wrote:

> The problem of the present section is therefore solved. Pure mathematics, as synthetical cognition *a priori* is possible only by referring to no other objects than those of the senses.

Kant opened Remark I of Part One by writing:

> Pure mathematics, and especially pure geometry, can have objective reality only on condition that they refer merely to objects of sense.

And he opened Remark II by writing:

> Whatever is given us as object must be given us in intuition. All our intuition, however, takes place by means of the senses only;

Given the difficulty of Kantian philosophy, it may not be too illuminating to claim that the above quotations from Kant express exactly what a strict finitist would say about mathematical objects. But I shall say it because it is true and because it suggestst the possibility of understanding Wittgenstein by comparison with Kant or Kant by comparison with Wittgenstein. For instance, my fumblings with the concepts of perspicuity, memorability, and surveyability are attempts to replace Kant's doctrines on space and time as an account of how we can determine the universal from what we see in the singular. However, I can add a few comments to show the plausibility of saying that a strict finitist would concur with these quotations from Kant. A strict finitist would agree that geometry can have objective reality, i.e., exist as a subject for human thought, only by referring merely to observable objects. He would agree because if the triangles referred to in geometry were unobservable REAL triangles, truths about them could not be established by looking only at the triangular shapes in his proof-patterns. In other words, if a strict finitist is right on how to establish necessity, geometry could not exist as a set of non-analytic necessary truths if geometry were about special mathematical objects rather than about the observable shapes of observable individuals. Yet these observable shapes are what some call only approximate triangles, approximate circles, etc.

But for Kant and strict finitists the triangles which we see are the real ones.

Also nothing that I have said so far shows that a strict finitist could not agree with Kant in holding that the only intuition is senuous intuition. However, the problem is not whether a philosopher wants to accept only senuous intuition. The problem is whether or not he can so restrict himself. In our discussion of strict finitistic epistemology, I shall argue that a strict finitist does not need a special non-senuous form of intuition for intuiting mathematical truth. But before this, let us compare strict finitism's views on mathematical objects with Berkeley's.

29. In sections 11 to 17 of the Introduction and 188 to 133 in his *Principles of Human Knowledge* Berkeley advances a philosophy of mathematics which for a minor exception is quite compatible with strict finitism. For instance in section 122 Berkeley wrote:

> those things which pass for abstract truths and theorems concerning numbers, are in reality conversant about no objects distinct from particular numerable things; except only names and characters, which originally came to be considered on no other account but their being *signs*, or capable to represent aptly whatever particular things men had need to compute.

(This quotation, and the other one, from Berkeley are from *Berkeley, Essay, Principles, Dialogues,* ed. by Mary W. Calkins, Charles Scribner's Sons, 1929.) A strict finitist would agree that arithmetic theorems need no objects besides particular numerable things. A strict finitist would agree that arithmetic theorems are conversant about words or signs in the sense that alteration of language is a consequence, – usually long range consequence –, of theorems. However, he would reject any suggestion that they are conversant about language in the sense that linguistic facts provide truth conditions for theorems. A proof leads a strict finitist to assert a theorem, not something outside the proof such as usage. And the strict finitist is right in doing so; we do not consider what people say as relevant to mathematical truth. As Wittgenstein put it in III-35:

Mathematical propositions seem to treat neither of signs nor of human beings, and therefore they *do* not.

Later in III-35 he repeated:

Once more: we do not look at the mathematical proposition as a proposition dealing with signs, and hence it *is* not that.

Indeed it can be misleading even to agree that mathematical propositions deal with signs in the sense that they determine connections in language. This can mislead because it suggests that the sole function of mathematical propositions is to express what we have proved and thereby cause an alteration in language. However, we also use mathematical propositions to do what Wittgenstein called predict, or what I would call compute the consequences of some data. Mathematical propositions can be used to compute predictions from empirical propositions. For instance, if we have an OVER or two apples and three pears, we can use our proved propositions to predict what the reversal will look like or how many pieces of fruit we will have. "Calculate" would be a good word to use here, but it has been pre-empted to stand for determining necessity rather than using necessary truths. "Predict" is suitable as long as it is not interpreted as saying that proved propositions are empirical predictions. Wittgenstein acknowledged that laying down a track in language is not the sole function of mathematical propositions when he wrote in III-18:

It is clear that mathematics as a technique for transforming signs for the purpose of prediction has nothing to do with grammar.

30. A strict finitist and Berkeley would, however, be in close agreement on how we determine necessity. At least they would agree that we can determine what holds of all figures of a certain kind by looking at a particular figure of that kind. A strict finitist would find nothing to quarrel with in Berkeley's statement in section 15 of Berkeley's Introduction to his *Principles of Human Knowledge*. It reads:

Thus, when I demonstrate any proposition concerning traingles, it is supposed that I have in view the universal idea of a triangle: which ought not to be understood as if I could frame an idea of a triangle which was neither equilateral, nor scalenon, nor equicrural; but only that the particular triangle I consider, whether of this or that sort it matters not, doth equally stand for and represent all rectilinear triangles whatsoever, and is in that sense universal.

A strict finitist would also go on to agree with Berkeley that to use a particular figure as a representative of a kind we must think abstractly about the particular figure without having any abstract image or figure before us. However, unless one can prove that developing philosophical theories is always pernicious, one should also try to give an account of what it is for an instance of an universal to represent any instance of that universal. The discussions of memorability, surveyability, and especially the discussion of perspicuity in section 23 were attempts to give such an account.

31. There are several ways of characterizing what we do in non-applied mathematics. As individuals, we determine what must be the result of applying certain operations to certain ordinary objects. For us, as individuals, our determination of what must be the result of applying these operations alters our concept of what these operations are. Or, in many cases, you could say that we have formed our concept of what it is to have applied the operation. Before the proof we have a concept of applying the operation while after the proof we have a concept of what it is to have applied it. But after the proof the concept of having applied the operation replaces the concept of applying the operation because in the proof-pattern we see the applying of the operation as being identical with having applied the operation. You could even say that as a result of the proof we change our concept of the operation from a temporal to a non-temporal concept. For instance, we changed our temporal concept of reversing the letters of OVER to the non-temporal concept of the reversal of the letters of OVER. Evidence that we have changed our concept for the temporal "reversing the letters of OVER" to the non-temporal "the reversal of the letters of OVER" is provided by the fact that we use the reversal for the criterion of whether or not we have

applied the operation. Mention of concepts of non-temporal connections suggests that we can characterize strict finitistic proofs as establishing the existence of internal relations. As a matter of fact Wittgenstein concluded *RFM* by talking about internal relations. The last paragraph of *RFM* reads:

> we can say that proof must shew the existence of an internal relation. For the internal relation is the operation producing one structure from another, seen as equivalent to the picture of the transition itself – so that now the transition according to this series of configurations is *eo ipso* a transition according to those rules for operating.

In I-99 to I-104, Wittgenstein indicated that an internal relation is asserted to exist when we speak non-temporally about the relation, i.e., speak as if relating the terms as the relation does is what it is to be the relation. For instance, being the reverse of is regarded as an internal relation between OVER and REVO when we identify being the reverse of OVER with REVO. And we assert that the relation is internal when we assert "The reverse of OVER is REVO" where the "is" is the non-temporal "is" of identity.

The concluding paragraph of *RFM* deserves a bit more comment. In this comment I shall talk about the proof of "The reverse of OVER is REVO" rather than the proof $2 + 3 = 5$ because I think that this sentence about OVER and its proof are the best examples to use whenever a difficult point in strict finitism must be discussed. First think about reversing the letters of OVER as a temporal transition. Think about it as the physical process or transition of producing another structure, usually a REVO, from OVER structures according to the rule: put the last letter first, the second to last second, etc. What is the criterion of whether or not someone has carried out this transition according to the rules? Stage 3 of Picture 2 provides the criterion, but how? Stage 3 is certainly a picture of such a transition. But Stage 3 of Picture 2 is taken as more than a picture of such a transition. Because an OVER is actually reversed in Stage 3 we take Stage 3 as being (equivalent to) a typical transition of this type. Once we have taken the picture of the transition as being a typical transition, we cease, for awhile, to think of the transition as a temporal process. The picture is static; it shows us before and after at the same time. By taking the transition out of time, we can survey it. The survey of Stage 3 leads us to identify reversing OVER with having a REVO. So now the series of configurations in Stage 3 of Picture 2, namely, the letters of the REVO being connected with the letters of the OVER, provides the criterion for whether or not one has correctly performed the temporal physical transition of having taken an OVER and rearranged its letters according to the rule: put its last letter first, etc. Of course, the criterion which the series of configurations provides is that the transition is correctly performed if and only if a REVO is produced. The series of configurations, viz., Stage 3, has provided us with this criterion because a survey of it has shown us that getting a REVO is what this transition is: its essence. To get the essence of the transition we had to look at it as an internal relation, i.e., a non-temporal relation or transition.

But, and this is the important point, no matter how we characterize what we do in a strict finitistic proof, we are talking about no individuals, properties, or operations other than ordinary observable ones. Even when we characterize what we are doing as determining and asserting the existence of internal relations, we are still dealing only with empirical relations and terms for these relations. To say here that the relations and terms are empirical is to say that we can produce observable instances of these relations and terms, which cannot be called mere approximations, to real instances of these terms and relations.

32. It would not be inappropriate for a strict finitist to join with Wittgenstein in II-29, where he wrote: "The limit of the empirical is concept-formation." A strict finitist would agree that in a proof we alter our concepts of empirical operations and objects. We make the alterations because we cannot see how anything other than what we have proved can be. Thus as a result of the proof we specify that certain results are not empirical possibilities; it does not make sense to say that they occur. For instance, it does not make sense to say that the reversal of OVER is REOV since we cannot have an idea of what it would be like

to reverse an OVER and get a REOV. So in proofs, concept-formation, we set limits on what are empirical possibilities, on what can be a fact. We set these limits because we do not have any clear idea of what it would be like for our proved propositions to be false empirical propositions.

Wittgenstein recognized the tremendous importance of not being able to conceive of a proposition being false. (See III-29 but especially III-4.) Actually one can look at much of *RFM* as an attempt to find the proper role of this much criticized psychological criterion for necessary truth. I shall discuss this inconceivability of the denial criterion for necessary truth in sections 39, 40, and 41. Here, I shall say only that I believe that Wittgenstein's showing us the proper place of this psychological fact is the most valuable contribution of *RFM*.

In brief, then: In non-applied mathematics we determine what we can think and what we cannot think about ordinary objects. Our proved propositions are expressions of what we as individuals have determined must be or cannot be. The long range linguistic consequence of our proofs is, that what we as individual provers have said cannot be thought and ultimately cannot be said about ordinary objects or anything else. So at least his mathematics does not require a strict finitist to accept anything into his ontology but that which is necessary for there to be ordinary objects, a language containing terms referring to ordinary objects, and people who can determine what can and cannot be thought about ordinary objects. But how can people determine this? This question brings us to epistemology, or strict finitism's answers to the questions in Group II. An epistemology is essential to the strict finitist since his ontology is a bonus from his epistemology. As usual though, his cutting of costs in ontology, viz., elimination of non-ordinary objects, forces him to pay a price in epistemology, viz., finding a place for self-evidence. But as we shall see it is not a high price.

33. The gist of strict finitism's epistemology is aptly summed up in the remarks that a mathematician is an inventor, as Wittgenstein said in I-167, and that what he invents is essence, as Wittgenstein noted in I-32. For instance, a mathematician invents the connection that makes the counting together of two and three

THE SAME as counting five. Previous to his invention of this identity, or creation of essence for counting together of two and three, the counting together of two and three merely almost always gave five but it was not identical with counting five. The main difficulty in strict finitistic epistemology is to provide an interpretation for "invent" or "create" so that the conventionalist view that any sentence we wish can be made into a device for expressing necessary truths is avoided. However, a strict finitist must not so sharply distinguish his position from conventionalism that he drifts into talking as if there were a kind of intuition which gave insight into pre-existing necessary connections. I hope I explain how a strict finitist this middle grounds when I develop his answers to the second group of questions.

34. A strict finitist should answer (IIi) by admitting that there are truths in non-applied mathematics, and he should go on to admit that these truths are necessarily true. He should make these admissions in order to avoid propounding the paradoxes that there are no mathematical truths and that everyone else is wrong in thinking that they must be true. However, a strict finitist should be reluctant to make these admissions without qualification. He should say that a mathematical proposition is true in the sense that the same set of words which are used to express it also expressed an experimentally justified truth. He should add that such propositions are necessarily true in the sense that we have proofs that they are unfalsifiable. But he should go on to say that the necessary truths of mathematics are not truths in the sense in which their empirical variants are truths; they do not have truth conditions. (The empirical variant of a strict finitistic mathematical proposition is the experimentally justified proposition expressed with the same sentence or set of words which are used to express it. For instance, the sentence "The number of letters in 'Bismarck' is 8" can be used to express the empirical truth justified by the kind of experiment pictured in Stage 3 of Picture 1. This empirical proposition expressed by "The number of letters of 'Bismarck' is 8" is the empirical variant of the proved proposition we finally express with the same sentence.) If he said that the mathematical truths had the same truth conditions as their em-

pirical variants, he, – a strict finitist –, would be saying that mathematical truths are equivalent to empirical truths. If he said that the mathematical truths have some other truth conditions, he most likely would end up saying that there are special mathematical objects. It would not help to say that the proof itself provides truth conditions for the mathematical truth proved by it. If a strict finitist said that the proof itself provided truth conditions for the proved proposition, he would be saying that those empirical conditions which are necessary for us to think the proved proposition unfalsifiable are its truth conditions. In particular, he would be in danger of saying that the contingent psychological fact that he cannot think of the proved proposition as being false provides the truth conditions for the necessary truth, – the proved proposition. (We shall see in our discussion of (IIiii) that a strict finitist does not claim that necessary truths assert the existence of those conditions which lead us to regard them as necessary.) In V-15 Wittgenstein made this very point that although empirical conditions are necessary to make the proved proposition's empirical variant true and psychological conditions are necessary for us to regard the empirical truth as unfalsifiable, the role of the proved proposition is not to assert the presence of these necessary conditions for its use as a necessary truth. A strict finitist would not be unhappy with the formula that a mathematical truth is a new use of an empirical truth, – its empirical variant –, and hence the mathematical truth is a different proposition from the empirical truth. Its truth comes from its empirical employment while its mathematical character, – its necessity –, comes from regarding what its empirical variant says usually happens as a result of a process as being what the process is.

35. A strict finitist has to admit, also, that there are mathematical falsehoods since people talk about statements such as $2 + 3 = 6$ being false. However, a strict finitist would want to qualify this admission. For a strict finitist we have a mathematical proposition only when we regard an empirical truth as unfalsifiable on the basis of a proof. It is very unlikely that we would find much empirical evidence for the empirical variant of what are called mathematical falsehoods. Further, I suspect that such evi-

dence as there is, for say $2 + 3 = 6$, is not representable in a surveyable picture. So a strict finitist would claim that there almost certainly are not any empirical falsehoods which are proved. So for a strict finitist, talk about mathematical falsehoods cannot mean that mathematics contains false propositions, i.e., that false propositions are proved. To talk about mathematical falsehoods must mean, then, to talk about a proposition which if an empirical truth would require us to think as empirically possible what we previously determined to be inconceivable, i.e., a proposition which conflicts with a previously proved proposition. So in a way mathematical falsehoods are mathematical only in a derivative sense. They are empirical falsehoods that have never been directly made into a mathematical proposition VIA a proof. But for every mathematical falsehood Q there is a proposition P, whose truth conditions we cannot imagine not occurring, but which if they failed to occur would make the falsehood Q empirically true, and P has been proved. Because of the way in which mathematical rules evolve, what I have just said is that a proposition can be regarded as a mathematical falsehood only by regarding it as conflicting with current mathematical rules. Without the rules it would not be mathematical. However, mathematical truths do not need rules to be made mathematical, i.e., to be proved. The difference between the true-false dichotomy as it applies to empirical propositions and as it applies to mathematical propositions is that there must be a mathematical truth before there can be a mathematical falsehood, while there does not have to be an empirical truth before an empirical falsehood is an empirical proposition.

36. How would a strict finitist answer question (IIiii) on what it means to say a truth is necessarily true in addition to being merely true and how we come to know that a truth must be true? To answer I must say that what it means to say that a proposition P is necessarily true. To say that P is necessarily true is to say that either one or two below holds.

One: P can be paraphrased as "The X of $Y_1, Y_2, \ldots\ldots, Y_n$, is Z" where "is" means identity.

Two: If P is rewritten in logical notation in which it has a form

Q, the proposition "The truth value assigned to Q in two value logic is true" is an identity.

If we say "White is lighter than black", "No red surface is green", and "All OVERs give REVOs when reversed" are necessarily true, we are saying that they meet condition One. We are saying that "The relation of shades which holds between white and black is white being lighter than black", "The number of surfaces which are uniformly red and green is zero", and "The reverse of OVER is REVO" are identities. I hope that these three examples show what I mean by "paraphrase" in condition One; we simply say the same thing except that we begin with "The X of". I talk about the paraphrasing simply because it enables me to call all assertions that a proposition is necessarily true assertions that a certain identity holds. A typical example of a proposition being necessarily true because it meets condition Two is provided by "This page is white or this page is not white." Since in logical notation the form of this proposition is (P v∼P), to say that the proposition is necessarily true is to say that "The truth value assigned to (P v∼P) is true" is an identity.

Now the above explication of what it means to say that a truth is necessarily true is not contained in *RFM*. However, I do believe that this explication is compatible with his philosophy of necessary truth, although maybe not compatible with his style of doing philosophy. In III-36 Wittgenstein wrote:

> That is to say: realizing mathematical relations has a role similar to that of realizing identity. It might almost be said to be a more complicated kind of identity.

37. The value of this explication of what it means to say that a truth is necessary is that it helps me explicate what is asserted when a necessary truth is asserted to be necessarily true. I can now say that if P is necessarily true, is asserted to be necessarily true, and meets condition One, then what is asserted is that the identity in the paraphrase holds. For instance, someone who asserts "2 + 3 = 5", where use of "=" has the force of asserting that 2 + 3 must be 5, asserts that the counting together of two and three is identical with the counting of five. In this example,

counting together of two and three and counting five are what could be called the topics of the assertion. If reversing the letters of OVER is our topic, someone who asserts, as a necessary truth, that the reverse of OVER is REVO asserts that "the reverse of OVER" and "REVO" stand for one and the same thing. So if a necessary truth is of type One, viz., meets condition One, to assert that it is necessary is to assert that an identity holds. If P is of type Two and is asserted to be necessarily true, even if P is all that is explicitly asserted, this is equivalent to asserting the identity: "The truth value assigned to the logical form of P is true." If there were not such an identity, then the assertion of P as a necessary truth of type Two is not acceptable, while if there is such an identity the assertion of P as a necessary truth of type Two is acceptable. All that I have said is that to assert that P is a necessary truth when P is necessarily true because it has a logically valid form is to assert that P has a logically valid form. But the important point is that even when we assert that an analytic truth is necessarily true we are asserting the existence of an identity.

38. The purpose of this explication is to enable me to distinguish between what a necessary truth means, viz., asserts when asserted to be necessarily true, and the grounds for it being a necessary truth. In particular, although a necessary condition for us to have grounds for saying that a proposition is necessarily true will be that we cannot conceive of its denial being true, what it means to say that a proposition is necessarily true will not be that we cannot imagine its denial being true. To repeat: to say that a proposition is necessarily true will be an assertion about an identity; not a psychological assertion. If a necessary truth is asserted but not asserted to be necessarily true, what is asserted is whatever it is that the sentence used to make the assertion says. So a strict finitist cannot be charged with the absurdity of trying to reduce necessary truths to contingent psychological truths about what we can and cannot conceive or imagine. But, still, it may seem that a strict finitist reasons in the following poor way. I cannot think of the X of Y not being Z. So the X of Y is identical with Z. To see that a strict finitist does not advocate such poor logic, we must develop a strict finitistic answer to the second part

of (II-iii). This part asks how we determine that a truth is necessarily true.

39. One of the best passages to keep in mind when considering how a strict finitist determines what must be true is the one from II-42 which reads: "It is not something behind the proof, but the proof that proves." In this context, this passage tells us that there is not some inference rule that tells us we can go from the strict finitist proof to the proved proposition. For a strict finitist the proof is the grounds for determining that a truth is necessarily true in addition to being merely true. There is no further justification beyond the proof. In particular, a strict finitist does not hold that a principle such as: "What we cannot imagine cannot be" justifies his proof. Still, because we cannot imagine a proposition being false we make it into a necessary truth. However, the "because" in the preceding sentence indicates a causal connection and not a logical inference. (This last point will be discussed shortly.)

If we regard proved propositions as statements on what are good inferences, we can interpret a remark of Wittgenstein's in V-45 as expressing a strict finitist position on what justifies establishment of a truth as necessarily true. In V-45 he wrote:

> Logical inference is a transition that is justified if it follows a particular paradigm, and whose rightness is not dependent on anything else.

Of course, it is the proof-patterns which are the paradigms which justify logical inference. Hence, logical inference will not justify moves from the proof-patterns. So if one does not think that a strict finitist has properly justified a claim that a truth must be true, the strict finitist is subject to challenge only by challenging his proof. Talk about logical fallacies will be irrelevant since his proofs are more fundamental than rules of logic. The best way to challenge a strict finitistic proof is to present a picture of what it is like for what is claimed must be true to be false.

But, although, a strict finitist does not seek a rule which permits him to go from "It cannot be imagined" to "It cannot be", doesn't he make such an unjustifiable move? The answer is: no, because the psychological fact of our being unable to imagine a proposition being false is not taken as ground for its truth or necessary truth. As Wittgenstein remarked in III-3:

> The fact, that is, that here we do not make experiments, but accept the self-evidence, is enough to fix the employment. For we are not so naïf as to make the self-evidence count in place of expirement.
>
> It is not our finding the proposition self-evidently true, but our making the self-evidence count, that makes it into a mathematical proposition.

(In the quotation, "self-evident" means cannot be imagined to be false.) But what does "making the self-evidence count" mean? It means that the prover responds to the self-evidence by treating the proposition about observable objects which he cannot imagine as falsifiable as a proposition which is not falsifiable by any experience. Part of such treatment is to assert that it must be true and to use it as a rule of inference to compute, prior to experience, what must be the result of certain observable operations or what must be the observable relations between observable objects. It would not be misleading to say that for a strict finitist the proof plays the epistemic role in the determination of what must be while the self-evidence is the cause of the proof's epistemic power. It could be misleading, though, to call the self-evidence "intuition" since the term intuition suggests that the self-evidence plays a justificatory role in the proof rather than a causal role because "intuition" suggests that there is an insight into the order and connection of non-ordinary objects.

40. At this point we can appreciate Wittgenstein's resolution to avoid developing philosophical theories to solve philosophical problems. In this case the philosophical problem is concern about the legitimacy of strict finitistic proofs since our previous beliefs about good reasoning tell us that a move from "is" to "must" is not justified. But strict finitists seem to make such a move. One way to solve this problem would be to assume that there is a special power of intuition which can see how properties MUST be connected from the way that they ARE connected in a perspicuous, memorable, and surveyable picture. Wittgenstein's way is to get us to look at strict finitistic proofs so that we will realize that it is

the strict finitistic proof that proves. He calls to our attention the fact that we do make the move from "is" to "must" and then we realize, contrary to previous beliefs, that in the presence of a proof such a move is not unjustified. He also acknowledges the fact that being unable to imagine the denial of the proved proposition, viz., its self-evidence, is a causal factor in our making the move. But as a philosopher he talks of self-evidence only to put it into perspective; he does not refer to it to give a philosophical explanation, i.e., justification, of any of our epistemic procedures. He simply describes the fact that self-evidence is a causal factor in our treating a proposition as unfalsifiable. So he reminds us that justifications, here proofs, are not justified by some other justification. We are reminded that if we doubt a justificatory technique, we should simply go back to the justification. If it works, our doubts vanish unless we have some irremovable preconceived idea about what a justification should be like regardless of what does or does not convince us. Of course, if we take a close look at the justificatory procedure and its use does not convince us of anything it is no good, and again the philosophical problem is solved. So, here, one of the main benefits of Wittgenstein's style of philosophizing by eschewing philosophical explanations is that we do not ignore the justificatory force of actual proofs by devising a theory to justify the justification actually used or on occasion to deny the justificatory power of justifications actually used.

41. When I began to develop strict finitism's answer to question (IIiii), I mentioned that a strict finitist should try to steer a middle course between intuitionism and conventionalism. We have just seen how strict finitism avoids intuitionism, where "intuitionism" means hypostatizing a special non-sensuous form of perception for recognizing necessary truths. To discuss strict finitism and conventionalism, I must say more exactly what conventionalism is. Michael Dummett in his "Wittgenstein's Philosophy of Mathematics", reprinted in *BP*, characterizes conventionalism on p. 494 of *BP*.

According to conventionalism, all necessity is imposed by us not on reality, but upon our language; a statement is necessary by vitue of our having chosen not to count anything as falsifying it. Our recognition of logical necessity thus becomes a particular case of our knowledge of our own intentions.

Dummett goes on to point out the inconsistency of a conventionalistic account of necessity which holds that all necessary truths are either our axioms, which are established by convention, or that which follows non-conventionally from our axioms. But not only has Quine in his "Truth by Convention", reprinted on pp. 322-445 of *BP*, noted the inconsistency of such conventionalism. In passages such as I-112 to I-121, Wittgenstein revealed that he was aware of such an inconsistency. So if Wittgenstein and strict finitists are conventionalists, they will be consistent about it. Dummett acknowledges on p. 495 of *BP* that they will be consistent conventionalists if conventionalists at all. Dummett writes:

Wittgenstein goes in for full-blooded conventionalism; for him the logical necessity of any statement is always the *direct* expression of a linguistic convention. That a given expression is necessary consists always in our having expressly decided to treat that very statement as unassailable; it cannot rest on our having adopted certain other conventions which are found to involve our treating it so.

What Dummett writes is true because a strict finitist admits that there are no necessary truths until people make what a sentence says into a necessary truth, i.e., until people treat what is said as unassailable. Also a strict finitist agrees that there is, so to speak, no necessary connection between necessary truths. But this passage is misleading, as is shown by the conclusions which Dummett draws. For instance, on p. 500 of *BP* Dummett writes:

He appears to hold that it is up to us to decide to regard any statement we happen to pick on as holding necessarily if we choose to do so.

On p. 501 of *BP* Dummett writes:

Wittgenstein's quite different idea, that one has the right simply to *lay down* that the assertion of a statement of a given form is to be regarded as always justified without regard to the use that has already been given to the words contained in the statement, seems to me mistaken.

42. What is wrong with Dummett's conclusions that a strict finitist and Wittgenstein hold that we can make any statement into a necessary truth? Logically any statement, even a contradiction, can be made into a necessary truth if we can find a proof for it. (See III-58, III-59, and V-21.) But strict finitists do not hold that we can make any statement a necessary truth merely by SAYING it is necessarily true or by LAYING DOWN that it is necessarily true. A proper appreciation of the role of proof in strict finitism and in Wittgenstein's thought will show that a convention is not sufficient to make a statement into a mathematical truth just as a proper appreciation of the role of proof showed that intuition was not epistemically necessary for a strict finitistic proof. To make a statement into a necessary truth requires a proof. So whether or not a strict finitist can lay down a statement as necessarily true depends upon the contingent fact of whether or not he can get a proof of it. If a man is not convinced that what he says must come out will always come out, he cannot with a clear conscience say that it must come out. More importantly, though, if he is not convinced that it must be as he says it must, and others are not either, one man saying it must be true will not produce a necessary truth in language.

Of course, if many people say that a truth such as "Twelve inches make a foot" must be true, even when they have no proof of it, it becomes an unassailable truth. But we can still distinguish it from truths such as "The reverse of OVER is REVO" because we can see the proof for the statement about REVO and OVER. We can think of changing the rule about inches and a foot, but we cannot think of altering the rule we now have on what happens if we reverse an OVER.

43. Here again we can appreciate Wittgenstein's descriptive approach to philosophy. The philosophical problem to be resolved is wonder about how it can be admitted that what is necessarily true is determined by people without admitting the absurdity that any statement can be made necessarily true. The wonder is dis-

solved by calling our attention to the fact that what people determine must be true is a truth which they are convinced must be true. It is just a fact that people are not convinced that most truths must be true. So it simply does not happen that any statement is made into a necessary truth. And when we look at what is required for conviction: perspicuous, memorable, and surveyable pictures, we can rest assured that there is little danger of people making most truths into necessary truths. We can now understand why Wittgenstein emphasized the importance of conviction in proofs. For instance in II-39 he wrote:

> What convinces us – that is the proof: a configuration that does not convince us is not the proof even when it can be shown to exemplify the proved proposition.

For philosophic purposes the importance of conviction in proofs is that it shows us that there are limits on what can be proved; Wittgenstein was not suggesting that one can proceed as follows. Look at a proof, introspect to decide whether or not you are convinced, and then assert the proposition to be proved if you have the proper feeling of conviction.

44. So my conclusion on conventionalism in Wittgenstein and in strict finitism is in agreement with that of Benacerraf and Putnam on p. 26 of their introduction to *BP* where they write:

> Wittgenstein's latter position is somewhat like Quine's: Quine and Wittgenstein both have a desire to destroy "platonistic" accounts once and for all, and at the same time, a distaste (albeit an ambivalent one) for conventionalistic accounts.

On pp. 26-27 they go on:

> What is unique to Wittgenstein is a tendency to write, at least part of the time, in a way that *suggests* that "what we say" somehow determines the truths of mathematics. Interpreters who want to make Wittgenstein a conventionalist find such passages grist for their mill; but it would probably be more reasonable (though perhaps not very charitable) to see this as an unfortunate way of writing produced by Wittgenstein's extreme antipathy to universals, necessary connections etc.

Of course, I do not agree with the last sentence in this quotation; I hope that I am showing in this essay how Wittgenstein talked about necessary connections.

45. A strict finitist can give satisfactory answers to the questions in Group III, viz., the questions about applied mathematics. Applied mathematics is the use of the inference rules which have developed or evolved from mathematical proofs. For instance, after we have a proof of "The counting together of two and three is counting five", we get $2 + 3 = 5$ as a rule, and we apply this rule when we conclude that we have five children because there are two girls and three boys. We could put it this way: In non-applied mathematics we determine, because of a proof, that a proposition which originally expresses an empirical truth that the referents of two terms are only always connected now says that the referents are identical. In applied mathematics we make use of the identity of referents established in pure or non-applied mathematics. So we could say that applied mathematics is differentiated from non-applied in that in non-applied mathematics we make truths necessarily true but do not talk about any objects while in applied mathematics we use these necessary truths and hence use the referring terms in them. As a result of using the referring terms in the necessary truths we do talk about objects; but these objects talked about are ordinary objects. They are ordinary objects because the terms in the necessary truth referred to ordinary objects before the truth was made necessary and the making of the truth into a necessary truth in no way altered the reference of the descriptive terms. After we proved that the reverse of OVER is REVO, "OVER" and "REVO" stood for what they stood for before the proof. (At most, in a strict finitistic proof the meaning of "is" is changed from the "is" of predication to the "is" of identity.)

In IV-41 Wittgenstein wrote: "Concepts which occur in 'necessary' propositions must also occur and have a meaning in non-necessary ones." This calls our attention to the fact that in non-applied mathematics we need a statement which is used to express an empirical truth as our candidate for a theorem. An instance of this statement must describe our proof-pattern. So the terms which occur in the theorem, – the necessary truth –, had to be used in a referring way before the proof.

So, if a strict finitist is to be criticized, he cannot be criticized by charging that he cannot give satisfactory answers to the three groups of questions on the philosophic foundations of mathematics. As a matter of fact, his simple non-platonistic and non-conventionalistic answers here are the best arguments for strict finitism. If strict finitism is wrong, it is wrong on the assumption that all mathematics can be understood as sufficiently like "The number of letters in 'Bismarck' is 8" for philosophical purposes. Or the strict finitists could be wrong because, although their philosophy of mathematics is desirable, it is part of an objectionable philosophy. We shall turn to a critique of strict finitism in the last chapter. In the next chapter, I hope to support my interpretation of Wittgenstein as a strict finitist by showing how several of his problems and positions can be understood if we read him as a strict finitist.

46. I shall close this chapter by arguing that Wittgenstein's use of "synthetic a priori" in III-42, where he wrote "The distribution of primes would be an ideal example of what could be called synthetic *a priori*, for one can say that it is at any rate not discoverable by an analysis of the concept of a prime number" is the same as the use of "synthetic a priori" in the opening quotation. I believe that the point of this passage from III-42 is that we cannot tell from any general rule that a number is prime although the definition of "prime" tells us what it means to say that a number is prime. In particular, the definition of "prime" does not give us a calculating procedure for determining "The prime number after Y is Z." For a given Y, we have to have a proof that a particular number is the requisite Z. For instance, the definition of "prime" does not tell us that the prime after 5 is 7. To determine this we need a proof that 7 is prime and a proof that there is no prime between 5 and 7. For very large primes it may be hard to see how there could be strict finitistic proofs. But there should be no difficulty in seeing how one could make a strict finitistic proof for "The divisors of 7 are 1 and 7". But the issue here is: How is "The prime after Y is Z" like "The

number of letters in 'Bismarck' is 8"? They are alike in that neither have the form of analytic sentences, both can be proved, – made necessary –, and if Y is sufficiently small both can be proved by strict finitistic means.

III

WITTGENSTEIN AS A STRICT FINITIST

> To resolve these philosophical problems one has to compare things which it has never seriously occurred to anyone to compare.
>
> In this field one can ask all sorts of things which, while they belong to the topic, still do not lead through its centre.
>
> A particular series of questions leads through the centre and out into the open. The rest gets answered incidentally.
>
> It is enormously difficult to find the path through the centre.
>
> It goes *via new* examples and comparisons. The hackneyed ones don't shew us it. (From V-12)

1. The opening quotation is from a discussion on what to say and what to do about a calculus in which a contradiction has been discovered. However, since the actual result of the discovery of the antinomies in set theory has been intense investigation into the mathematical and philosophical foundations of mathematics, it is not perverse to interpret the above passage as a comment on the entire philosophy of mathematics. In my opinion what Wittgenstein thought should be compared is mathematical propositions with propositions such as: "The number of letters in 'Bismarck' is 8" and "The reverse of OVER is REVO" and mathematical proofs with strict finitistic proofs of such propositions. The sort of questions which do not lead to the center of the problem are the sort of questions which lead us to do more mathematics while trying to philosophize about mathematics. Such peripheral questions would be questions such as how to establish the consistency of a calculus or how to re-write mathematics without quantifying

over predicate terms. Despite the difficulty in interpreting Wittgenstein, I shall go out on a limb and say that the three groups of questions asked about strict finitistic proofs after mathematical proofs have been compared with them, – construed as them –, is the series of questions which leads through the center and out into the open, viz., sets one on the way towards developing a satisfactory philosophy of mathematics. I shall not, however, argue for the strong point that Wittgenstein was perfectly satisfied with his strict finitism. I am arguing only that he takes it as his starting point. He even admitted in Appendix II-18 that he had only indicated how a complete investigation would go. From the perspective of my interpretation, answering the three groups of questions by developing a metaphysical theory or by talking about mathematics as a formal calculus is making use of hackneyed examples and comparisons. Unfortunately for my interpretation, Wittgenstein did not explicitly ask or answer these questions or any questions like them. But I can explain away his not explicitly answering these questions by pointing out that this book, – *RFM* –, is a selection of notes and not an organized treatise.

In this chapter I propose to support my interpretation of Wittgenstein as a strict finitist in the following way. I shall show that we can understand several of his positions on mathematics and problems which he has about mathematics if we read him as being a strict finitist. These positions and problems are as follows. He wondered whether or not Gödel's Incompleteness Theorem, or any *reductio ad absurdum* proof, makes sense. He took the position that a mathematical proof gives meaning to the proved proposition. He had the problem, which is related to the problem that proofs give meaning to the proved proposition, of explaining how one proposition could have two proofs. He held that the law of excluded middle did not have unlimited application. He doubted the legitimacy of Cantor's transfinite numbers and Dedekind's introduction of real numbers. He held the apparently outrageous position that there is nothing mathematically wrong with contradictions. Finally, he continually worried about what is our guide in the n-th term in a series.

2. Alan Ross Anderson argues that Wittgenstein's discussion

of Gödel's Incompleteness Theorem is incompetent. In his "Mathematics and the Language Game", *BP* 481-490, Anderson wrote of Wittgenstein's discussion of Gödel's Theorem: "But we are in point of clarity worse off after Wittgenstein's account than before." However, I believe that Wittgenstein's discussion of Gödel's Theorem is not as unfortunate as Anderson claims. On the contrary, Wittgenstein's discussion of Gödel's Theorem is philosophically clarifying because it calls our attention to a philosophical assumption behind Gödel's proof. The assumption is that a proposition can be a mathematical proposition, viz., have meaning as a mathematical proposition, independently of and prior to our having a proof of it. In the next two sections we shall consider doubts about this assumption and its role in Gödel's Theorem. But here, in fairness to Anderson, I should add that because of his resolution to account for existing mathematics, Wittgenstein would not conclude that Gödel's Theorem is bad mathematics. So Wittgenstein's discussion did suffer from the unclarity of inclusiveness.

3. In Appendix I, which is devoted to a discussion of Gödel's Incompleteness Theorem, he wrote in 15:

> Whether something is rightly called the proposition "X is unprovable" depends on how we prove this proposition. The proof alone shew what counts as the criterion of unprovability. The proof is part of the system of operations, of the game, in which the proposition is used, and shews us it 'sense'.

And in 16 he wrote:

> The proposition "P is unprovable" has different sense afterwards – from before it was proved.

In IV-42 Wittgenstein wrote:

> The problem of finding a mathematical decision of a theorem might with some justice be called the problem of giving mathematical sense to a formula.

A strict finitist also holds that the proof of a proposition changes its sense and that proof gives a mathematical sense to a proposition. A proof of a proposition such as "The sum of two and three is five" or "The reverse of OVER is REVO", changes the meaning of the proposition by changing the meanings of the concepts in the proposition. For instance, as a result of a proof we alter the concepts of "The sum of two and three" and "five" by determining that there is a necessary connection between them. Before the proof we were aware merely of an empirical connection of constant conjunction between that to which we applied "The sum of two and three" and that to which we applied "five". As a result of the proof, however, we change the meaning of "The sum of two and three" by determining that calling an aggregate the sum of two and three is the same as to apply "five" to it. So since the concepts of a proposition are altered, the proposition cannot have the same sense after the proof as before. Also a strict finitist holds that a proof gives mathematical sense to a proposition in that a proof is what leads us to determine that an experimentally justified proposition is a necessary truth. By making it into a necessary truth we make it into a mathematical proposition or at least make it into the kind of proposition which can be used to compute prior to experience what we can experience. Or we could put it this way: As a result of a proof we change the empirical "The X of Y is Z", where "is" is the "is" of predication to the mathematical "The X of Y is Z", where "is" is now the "is" of equality.

So we can see that if we assume that Wittgenstein were a strict finitist he would, as he did do, claim that proof alters the meaning of a proposition and gives it a mathematical sense.

4. We can also, now, see why he wrote in V-19:

> However, queer it sounds, my task as far as concerns Gödel's proof seems merely to consist in making clear what such a proposition as: "Suppose this could be proved" means in mathematics.

In Gödel's proof we suppose that an arithmetic proposition, P, which says of itself under normal interpretation that it cannot be proved by the techniques of a specific axiom system A, could be proved by the techniques of axiom system A. (Here to say that "P" is arithmetic and the statement "P is not provable by techniques of A" is arithmetic is only to say that they can be written using signs for numbers, number variables, plus, times, successor, equality, and quantifiers. It is not to admit that either P or the statement "P is not provable by the techniques of A" are either necessary truths or necessary falsehoods.) But what does "P is provable by the techniques of A" mean and what entitles us to call it a mathematical proposition? If we are strict finitists only a proof will fix the meaning of the concepts in it and make it mathematical; by "fix the meaning" I mean give the concepts the new use that they will have after the proof. For a strict finitist, it would be a proof that P is provable by the techniques of A which would show him what it is like for P to be proved by these techniques of axiom system A. So for a strict finitist to suppose that P is provable by the techniques of A is to suppose that he knows what it would be like for P to be so proved. The supposition that he knows what it is like for P to be proved by the techniques of A is the supposition that he knows what it is like to have a sequence of formulae constructed according to the rules for forming proof sequences in system A and whose end formula is P.

Now we have a dilemma. You can write down what you supposed you knew or you cannot. If you can write it down, then you exhibit a proof in A of P and Gödel's proof could not be given. (Incidently, such a proof sequence in A would provide a strict finitistic proof-pattern for "P is provable in A.") If you cannot write down what you supposed you knew, this is because you cannot think of what it would be like for P to be proved in A. If you cannot think of what it would be like for P to be proved in A, you cannot think of what it is like to have a strict finitistic proof of "P is provable by the techniques of A." To know what it is like to have a strict finitistic proof of a proposition is to have one. If we are strict finitists, and cannot think of what it is like to have a strict finitistic proof of "P is provable by the techniques of A", we cannot suppose "P is provable by the techniques of A" is a mathematical proposition. If it is not a mathematical proposition, how can we use it in a mathematical proof?

As can be seen from the above argument, strict finitistic worries about Gödel's Theorem are really worries about the legitimacy of *reductio ad absurdum*. I submit that Wittgenstein called his own treatment of Gödel's Theorem queer sounding because he was discussing this highly technical result without being concerned with any of the details of its techniques; he was discussing only the legitimacy of a feature of it, viz., being a *reductio*, which it shares with many unsophisticated and trivial proofs. This fact that Wittgenstein not only questioned the legitimacy of *reductio* in Gödel's Theorem together with the fact that in IV-28 he questioned its use in general gives support to the claim that he can be read as a strict finitist. A strict finitist would object to any use of *reductio ad absurdum* since the *reductio* assumption is not proved and hence is not a meaningful mathematical proposition. And if it is a meaningful mathematical proposition because its denial is proved, it is an unthinkable false mathematical proposition, and there is no need to try a *reductio* proof since we already have what would be proved by the *reductio*.

5. Of course, if a strict finitist claims that the proof gives meaning to the proved proposition, he faces the objection that one would not know before he completed his proof what he wanted to prove. In V-7 Wittgenstein considered such an objection. He wrote:

> Now how about this – ought I to say that the same sense can only have *one* proof? Or that when a proof is found the sense alters?
>
> Of course some people would oppose this and say; "Then the proof of a proposition cannot ever be found, for, if it has been found, it is no longer the proof of *this* proposition". But to say this is so far to say nothing at all. –

A strict finitist can dismiss this objection as summarily as Wittgenstein did. The function of a proof, for a strict finitist, is to give a sentence, – a verbal form for expressing propositions and making statements, – a new use. For instance, in the proof in Chapter II, section 13 we gave the sentence "The reverse of OVER is REVO" the new use of being a rule for computing a priori what the reverse of an OVER is. We gave it this new use by further fixing the meaning of, viz., giving new criteria of application for, the concepts "the reverse of OVER" and "REVO", in addition to making the "is" into equality. So what should we say happens as a result of a proof? A proposition P is given a new use? Or should we say that a proposition P is changed into a new proposition Q, – a mathematical one? It makes no difference what we say. The important thing is to understand what we do in a proof, not how we describe what we do in terms of, the as yet, unclear concept proposition. (Of course, I am not saying that we should describe what we do in a proof without use of the concept of "proposition". It is a useful concept; Wittgenstein used it frequently in *RFM*. Unfortunately, however, there are no generally accepted grounds for the identity of propositions. But this means only that we cannot answer "What is a proposition?" So, when we use "Proposition" we should try not to use it in contexts where it is necessary to say what one is and we should simply try to use it as others do.) I prefer to say that we convert an empirical proposition P into a mathematical proposition Q although I cannot see how this characterization is better than one which says that what we do is take a proposition P, which has an empirical use, and give P a mathematical use. A strict finitist would agree with how Wittgenstein continued V-7:

> It all depends *what* settles the sense of a proposition, what we choose to say settles its sense. The use of the signs must settle it; but what do we count as the use? –

Here Wittgenstein was claiming that it really is a matter of pure convention whether or not we say that the proposition we end up proving is the same as the one we started out to prove. He adds that it does not help to say that the use of a sentence will enable us to determine whether or not it expresses the same proposition, since there are no generally agreed upon grounds for identity of use. We simply are not clear about which uses of a sentence identify it as expressing the same proposition or a different one. I take this fact that strict finitists and Wittgenstein would face a similar problem and handle it in a similar way as further evidence in support of my suggestion that it is helpful to read Wittgenstein as if he were a strict finitist.

6. A strict finitist would hold that, in a sense, a proposition can

have only one proof. A proof of "The X of Y is Z" has a proof pattern consisting of a perspicuous instance of Y, a memorable picture of taking the X of that Y, and a picture in which we can see, i.e., survey to realize, that the X of that Y is identical with the perspicuous Z in the picture. (The proof also includes our appreciation of the fact that in the pattern the X of Y is identical with Z. But here I am concerned not with the interior aspects of proof, only with the public proof pattern. If "proof" means individual act of proving, a proof is never repeated.) This pattern, after the proof, is now our model for what it is to take the X of Y. Wittgenstein was in agreement with strict finitism when he wrote in II-24:

> The proof must be our model, our picture, of how these operations have *a result.*

and

> The proof is our model for a particular *result's being yielded,* which serves as an object of comparison (yardstick) for real changes.

Since taking the X of Y is one thing, there is only one model of what the operation is, although there may be many instances of this model. In the sense that there can be only one model of what an operation is, a strict finitist holds that a proposition can have only one proof, i.e., one type of proof-pattern. He would agree with II-59 in which Wittgenstein wrote:

> But this means e.g.: this proof is a mathematical entity that cannot be replaced by any other; one can say that it can convince us of something that nothing else can, and this can be given expression by our assigning to it a proposition that we do not assign to any other proof.

7. Of course, a strict finitist would have to admit what Wittgenstein admitted in II-58: "Of course, it would be nonsense to say that *one* proposition cannot have two proofs – for we do say just that." So a strict finitist, as well as Wittgenstein, has to show how his view of proof is compatible with the fact that we ordinarily talk about a proposition having two proofs. There are two distinct cases to consider. In the first case, we say that a proposition has

two proofs although it is admitted that the two proofs are merely variants of the same pattern. Such a case would be proving $2 + 3 = 5$ with $\underbrace{\overbrace{||}\;\overbrace{|||}}$ and $\underbrace{\overbrace{a\,a}\;\overbrace{b\,b\,b}}$ In the second case, we say that the same proposition has two distinct proofs: one in one system and another in another system. Examples of case two are: Proofs of a geometric proposition in synthetic and in analytic geometry, Proofs of $2 + 3 = 5$ from $\underbrace{\overbrace{||}\;\overbrace{|||}}$ and by construction of a derivation of it from some axioms as $1 + 0 = 1$, and Successor $(a + b) = a = +$ Successor (b). A third case occurs when the proposition is rewritten as a formula in a formalized axiom system and at least two distinct proof sequences have the formalized version of the proposition as their end formula.

How could a strict finitist handle these cases and would he handle them in the same way as Wittgenstein did? A strict finitist would handle the first case by saying that two distinct proof patterns, such as $\underbrace{\overbrace{||}\;\overbrace{|||}}$ and $\underbrace{\overbrace{a\,a}\;\overbrace{b\,b\,b}}$, are two instances of one and the same proof-pattern for a proof of "The X of Y is Z" if and only if they are both perspicuous for X and Z, memorable for "the X of Y", and we can survey them to see that the X of Y is Z. If they are not instances of the same proof-pattern they cannot be used to prove the same thing since we will not see the same things in them. In other words, two proof-patterns are the same if no one of them is better than the other as a model of what the X of Y is. If two proof-patterns are not models for the same thing, they are not proofs of the same proposition. Of course, a strict finitist would point out that two instances of the same proof-pattern can be called different proofs but this is not a significant sense of "two proofs"; it really would be better, in this case, to talk about instances of the same proof. Unfortunately for my interpretation of Wittgenstein, I cannot find any passage where Wittgenstein gives the same solution. However, the following remark from II-58 can be read as giving a strict finitist solution to case one of multiple proofs.

"Two proofs prove the same when what they convince me of is the same." – And when is what they convince me of the same? – How do

I know that what they convince me of is the same? Not of course by introspection.

I read the last three sentences as follows. Two proof-patterns convince us of the same proposition if they serve equally as well as models. Whether or not two proof-patterns serve equally well as models is not, of course, decided by introspection; it is decided by looking at the proof-patterns to see whether they are perspicuous, memorable, and surveyable in the same way. If they are, they are the same model; otherwise not.

A strict finitist would handle the second case in the following way. To say that the same proposition P is proved in two different systems S_1 and S_2, requires a proof that the two different proofs prove the same proposition. To fulfill this requirement a strict finitist would have to give a strict finitistic proof of a proposition of the following form. The translation of proposition Q, proved by proof P_1 in system S_1, into the language of S_2 is a proposition Q* proved by proof P_2 in S_2. An instance of this formula is: The translation of "Two straight lines have at most one intersection" proved by geometric means into analytic geometry is "The system of equations $ax + by = c$ and $dx + ey = f$ has at most one solution if not infinitely many" proved by algebraic means. So for a strict finitist a proposition having proofs in different systems is simply a bit of mathematics; a proposition has the two proofs if and only if there is the proof that it has the two proofs. The proof of the fact that the proposition has two proofs will show what it means for the separate proofs to give the same meaning to the proposition proved by both.

I submit that Wittgenstein would give the same account of what we have when we have proofs in different systems of the same proposition. Certainly a strict finitist would agree with II-62 where Wittgenstein wrote:

Let us remember that it is not enough that two proofs meet in the same propositional sign. For how do we know that this sign says the same thing both times? That *must* proceed from other connexions.

For instance, although different proofs may have as their conclusion that non-parallel straight lines have at most one point of intersection, we do not know that this form of words says the same thing, expresses the same conclusion, unless we have a proof, – other connections –, which show that the proofs prove the same thing. I believe that passages such as the following two, the first from II-46 and the second from V-7 conclusively establish that Wittgenstein's account would be the same as the strict finitists.

> Now it is possible to image some – or all – of the proof systems of present-day mathematics as having been co-ordinated in such a way with one system, say that of Russell. So that all proofs could be carried out in this system, even though in a roundabout way. So would there then be only the single system – no longer the many? – But then is must surely be possible to *shew* of the *one* system that it can be resolved into the many.

I italicized "shew" to stress that to say a proof in one system proves a proposition in another system requires proof. The next quotation from V-7 is also cited to establish that Wittgenstein held that it must be proved that proofs in different systems proved the same proposition.

> When two proofs prove the same proposition it is possible to imagine circumstances in which the whole surrounding connecting these proofs fell away, so that they stood naked and alone, and there were no cause to say that they had a common point, proved the same proposition.

For instance, we can imagine what it would be like to have no analytic geometry and hence for certain algebraic propositions there would be no geometric sense.

8. Also, if we read Wittgenstein as taking the strict finitist's approach towards accounting for proofs in different systems of the same proposition, we can reconcile an apparent inconsistency in his claim that proofs give meaning to proved propositions. In V-7 he also wrote:

> I once said: 'If you want to know what a mathematical proposition says, look at what its proof proves.' Now is there not both truth and falsehood in this? For is the sense, the point, of a mathematical proposition really clear as soon as we can follow the proof?

I submit that the element of falsehood is that we can have a proof of a proposition in one system but not yet fully understand the point of the proposition. We do not fully understand its point because we do not yet have a proof that a proof in another system proves it; hence we do not yet understand what the other proof will show us can be done with the proposition. For instance, simply proving by algebraic means that the system of equations: $ax + by = s$ and $dx + ey = f$, has at most one solution if not infinitely many does not give us the geometric point of this proposition.

9. However, my interpretation that Wittgenstein is a strict finitist is weakened by the fact that he did not explicitly claim that the proof that a proof in one system proves the same proposition as a proof in another system must be a strict finitistic proof. I think that his view on establishing the equivalence of systems was as follows. For purpose of illustration let the two systems be: Reference to strokes such as || ||| to calculate sums and use of ordinary calculating rules such as $9+3=12$, and $8+5=13$, etc., arithmetic as taught in elementary schools. In this case, we could say that the use of the strokes is the long way of giving a proof; this should be obvious for calculating a sum such as: The sum of 97 and 73 is □. We can call the ordinary calculating rules the decimal system. Up to a point where the stroke proofs get confused, the stroke proofs can justify what we say in the decimal notation. Consequently, the stroke proofs themselves are proofs that what is proved with the strokes is also proved in the ordinary decimal system of addition. The stroke proofs are, of course, strict finitistic proofs. So up to a point there is a strict finitistic proof of the equivalence of the stroke and the decimal system.

After awhile, however, we can no longer justify what we do in the decimal notation by reference to stroke patterns because the requisite stroke patterns would no longer be perspicuous, memorable, and surveyable. When this point is reached, what we do in the decimal system is no longer merely a different notation for what we can do with the strokes because now, as Wittgenstein pointed out in II-47, only use of the decimal notation determines truths of addition. It is as he suggested in II-51: Calculation in the decimal notation has a life of its own. We can now see why Wittgenstein wrote in II-45:

But how can one technique of proof be *defined* by means of another? How can one explain the *essence* of another? For if one is an 'abbreviation' of the other, it must surely be a *systematic* abbreviation. Proof is surely required that I can systematically shorten the long proofs and thus get once more a system of proofs.

Long proofs at first always go along with the short ones and as it were tutor them. But in the end they can no longer follow the short ones and these shew their independence.

We now have the disturbing result that we started out with a primary system, the strokes, which we used to justify a secondary system for doing the same kind of calculation, the decimal system. But after awhile we cease to be able to justify what we do in the secondary system by reference to the primary system. So how do we justify what we do later on in the secondary system? It does not suffice as proof simply to say that if need be we could always return to the primary system to justify what we do in the secondary system. We can have a vague idea of checking every addition in the decimal notation by reference to stroke patterns just as we can have a vague idea of doing every arithmetic calculation in the notation of logic. Still, possession of this vague idea does not show that we have a more ultimate method of proof at our disposal, because if we tried to produce one of the alleged more ultimate proofs it would be too complicated to prove anything. In regard to saying that we can justify calculating techniques by reference to a primary system, Wittgenstein remarked in II-53:

"But at every step in the secondary system, we can imagine that it could be taken in the primary one too!" – That is just it: *we can imagine that it could be done* – without doing it.

Of course, imagining that we have a proof is not having one or even being able to have one. Although a strict finitist would claim that a proof in the imagination is a proof, this is not to claim that imagining we can have a proof is a proof. (I should note that in the preceding quotation the primary system is doing arithmetic in the notation of logic. But this does not affect my point because proofs in logical notation are just as worthless as those in stroke notation for very large calculations.)

Wittgenstein went on in II-54 to argue, unclearly I feel, that there cannot be a strict finitistic proof, viz., a stroke system proof, that whatever is legitimately done in the decimal notation is justified by a stroke proof. I think that his point was that circularity would be involved in an attempt to prove that what is done in the decimal notation can be done in the stroke system. To prove this, we would need to prove that for every numeral there is a distinct pattern of strokes. To prove that for every numeral there is a distinct pattern of strokes requires mathematical induction. In this inductive proof we would say something like: If a numeral $n-1$ having associated with it a pattern with exactly $n-1$ strokes implies that numeral n has associated with it a pattern with exactly n strokes, then any numeral n has associated with it a pattern with exactly n strokes. Now to prove in the stroke system that what is done in the decimal system can be done in the stroke system requires that the above statement of induction be re-expressed as a statement about strokes. But such a re-expression requires that any statement about numerals can be re-expressed as a statement about strokes, and that is what is to be proved.

How, then, do we according to Wittgenstein, prove that we have a right to replace one system with another? He answered in V-7:

> If you ask: "What right have we?" the answer is: perhaps none – What right have you to say that the development of this system will always run parallel with that one?

So here is where I have to admit that I cannot conclusively establish that Wittgenstein is a strict finitist. He admitted that we can move to new systems of proof; but he did not require that this move be justified by a strict finitistic proof. Indeed, he did not require that the move be justified by any kind of mathematical proof. (I do believe, however, that the first stages of a new system would be justified by strict finitistic proofs, even according to Wittgenstein.) Since the move from one system to another is a significant mathematical move, I cannot say that Wittgenstein's account of all mathematics is strict finitistic. Here it seems to be conventionalistic. But this is not to admit that he gives a conventionalist account of any mathematics done within a system.

Fortunately for his account of mathematics, here again his descriptive approach to philosophy takes the oddity out of even this little bit of conventionalism in his philosophy of mathematics. He reminds us that simply because there is no extra-systematic justification for systems, it does not follow that any system is acceptable. I-4 contains such a reminder.

But is this counting only a *use*, then isn't there also some truth corresponding to this sequence? The *truth* is that counting has proved to pay.

But Wittgenstein did not want us to be misled to the view that the utility of a system is a ground for saying that it is true or consists of truths. So he continued I-4:

"Then do you want to say that 'being true' means: being usable (or useful)? – No, not that; but that it can't be said of the series of natural numbers – any more than of our language – that it is true, but: that *it is used*, and, above all, it is used.

A philosopher, like Wittgenstein, cites the fact that used systems are useful to remind us that as a matter of fact despite the lack of proof for systems as a whole, there are limits on what systems will be used. Thus worries about all kinds of insane calculating procedures being acceptable if we admit that convention only justifies use of a system are shown to be baseless worries. In fact people will not accept insane calculating procedures.

10. What would a strict finitist say about proofs in a formal system of a propositional form, such as the proofs in the propositional calculus of the formula: $(P \supset Q) \supset ((P.Q) \supset Q)$. When we consider such formulae and such systems, it seems preposterous to hold that a proposition can have only one proof. Wittgenstein noted this apparent absurdity when he began II-60 by remarking:

But am I not making a crude mistake? For just this is essential to the propositions of arithmetic and to the propositions of the Russellian logic: various proofs lead to them. Even: infinitely many proofs lead to any one of them.

In this case, I think that Wittgenstein's way of explaining away this apparent absurdity is the only course open to strict finitists. If we have a formalized axiom system, A, a formula F, which is a theorem of A, and sequences of formulae S_1 and S_2, each of which is a proof of F, Wittgenstein's account would be as follows. As long as the sequences S_1 and S_2 are not too complicated, they provide strict finitistic proof-pattern for propositions in the metalanguage of A of the form: The construction of sequence S by rules of A is a proof in A of F. Thus the many formal proof sequences leading to the same formula F do not provide many proofs for a proposition F. (F is, anyway, a formula, not a proposition.) They provide only many proof-patterns for many different metalinguistic propositions. If one wished, it would not be odd to call these metalinguistic propositions geometric propositions since they tell what new shapes can be constructed from some other shapes according to certain rules; they tell what new formulae can be constructed from some origanal formulae in A according to certain construction rules of system A.

In II-43, II-64, and III-14, there are remarks which show that Wittgenstein gave the preceding account of formal proofs. However, II-38 contains both a statement of this account plus a slight modification of it. So I shall quote from II-38 to establish that Wittgenstein gave the preceding account of what is proved by formal proofs. He began II-38 with:

> Accepting a proof: one may accept it as the paradigm of the pattern that arises when *these* rules are correctly applied to certain patterns.

Later on in this Remark he noted that if he held:

> "The proof proves first and foremost that this formation can arise when one operates with these signs according to these transformation rules."

he gives a geometrical sense or point to formal proofs. He wrote:

> For the proposition whose truth, as I say, is proved here, is a geometrical proposition – a proposition of grammar concerning signs.

As the following statements from II-38 show, he held, for the most part, that formal proofs have this role as proof-patterns for geometric propositions.

Or again: when mathematics is divested of all content, it would remain that certain signs can be constructed from others according to certain rules. –

The least that we have to accept would be: that these signs etc. etc. – and accepting this is a basis for accepting anything else.

I submit that what he means here by "a basis for accepting anything else" is that before we can maintain that a sentence is a necessary truth because it has the form of a theorem of a certain formal axiom system, we need the geometric, – strict finitistic –, guarantee that the form is really a theorem of the axiom system.

However, at the end of II-38, he modified his suggestion that geometric certainty that a formula is a theorem of a specified axiom system is necessary and sufficient to determine that propositions with that form are necessary truths. For instance, knowing that the move from (x) (Fx) to Fa is sanctioned by the rules of a system, or knowing that (Ey) ((x) (Fx) $\supset$ Fy) is a theorem, is neither necessary nor sufficient for us to know that: "If all bodies have weight, then my body has weight" must be true. We have to know whether the system is right. But this modification does not alter his account of what it is for a formula to have many proofs in a formal system. The modification shows only that strict finitistic proofs that a formula is provable in a system have only limited utility in establishing what is a logical truth; in particular they cannot establish that fundamental rules of logic such as *modus ponens* and universal instantiation are valid inference forms. It is not inconsistent with strict finitism to hold that formal proofs are not the sole proof-patterns for establishing that a formula is the form of a logical truth. In my "Deduction and Intuitive Induction", I tried to show how truth tables and Venn diagrams could be used as strict finitistic proof-patterns for fundamental logical truths.

11. Unfortunately for my interpretation of Wittgenstein as a strict finitist, he did not explicitly say in *RFM* that fundamental rules of logic such as *modus ponens* and universal instantiation are to be justified by strict finitistic means. As a matter of fact, early in *RFM*-I he wrote as if universal instantiation is a good inference simply because people make it. (See I-8 to I-17.) But he

is still bothered by suggesting that anything can follow from anything else if enough people draw the conclusion long enough. So from I-25 up through I-107 he discussed primarily how our determination of what must be depends upon pictures of how experiments almost always turn out. (I developed the account of how stict finitists would proceed as mathematicians from these Remarks.) I interpret him as going into the discussion of what I call strict finitistic mathematics in order to show how laws of logic can be non-arbitrary rules of language without corresponding to any facts. Unfortunately, after his discussion of strict finitism in *RFM*-I when he turns to a discussion of laws of logic, his primary concern still seems to be to show that logical truths are not true because they correspond with some kind of logical fact. He does not stress that certain pictures make the laws of logic non-arbitrary; that certain pictures force the laws of logic on us. However, I-131 and I-133 give some support to my interpretation that he would say that fundamental rules of logic should be justified by strict finitistic means. In I-133 he wrote:

> The propositions of logic are 'laws of thought', 'because they bring out the essence of human thinking' – to put it more correctly: because they bring out or shew, the essence, the technique of thinking.

This remark supports my interpretation inasmuch as in passages such as I-23 he claims that it is the job of a strict finitistic mathematician to create the essence of a thing. I think that his view in *PI*-371 that essence is expressed by grammar, i.e., the essence of a thing is given by what we must and cannot say of it, plus his view that grammatical connections are created and not discovered, entitles me to say that all essences are created. Since he never talked about essences being created by means other than strict finitistic means, I assume that he held that all essences are created by strict finitistic means. With these assumptions, I can conclude that the laws of logic, – the essence of human thinking –, are established by strict finitistic means in Wittgenstein's opinion.

12. What does a strict finitist say about the law of excluded middle and what did Wittgenstein say about it? This question is

ambiguous because it could be taken in three ways. It could be taken as asking what they would say about the propositional form (P v∼P) where "v" and "∼" have their standard interpretation. It could be taken as asking what they would say about: "Every declarative sentence in which there are no terms without referents is true or false." Thirdly, it could be taken as asking what they would say about: "Every sentence of the form (P v∼P), where there are no terms without referents in P, is true." We can discard the second version of the question because, as Wittgenstein noted in his *Tractatus* at 6.111, a poor way to state the law of excluded middle is to say that every proposition is true or false. His *Tractatus* statement reads:

> Theories which make a proposition of logic appear substantial are always false. One could e.g. believe that the words "true" and "false" signify two properties among other properties, and then it would appear as a remarkable fact the every proposition possesses one of these properties. This by no means appears self-evident, no more so than the proposition "All roses are either yellow or red" would sound even if it were true.

(Note that even at the *Tractatus* stage of his thought he wanted logical laws to be self-evident.) So when we ask for a philosopher's position on the law of excluded middle, we are asking what he would say about the tautologousness of (P v∼P) and what he would say about the truth of every sentence with the (P v∼P) form. These are two distinct questions, as we shall see in section 14. We shall see that one can claim that (P v∼P) is a tautologous form while not predicating "true" to propositions with that form.

13. A strict finitist would say that truth tables or diagrams such as those in *Tractatus* 6.1203 provide strict finitistic proof-patterns of the tautologousness of forms such as (P v∼P). Of course, even at the *Tractatus* stage of his thought Wittgenstein would give a strict finitistic proof of the tautologousness of (P v∼P). In *RFM* his use of the concept "tautology" in passages such as II-41 indicate that he did not change his concept of tautology from the *Tractatus*. Nor did he change his position that tautologousness is established by strict finitistic means, although in *RFM* he did not talk about establishing that a form was tau-

tologous by means of a truth table. In *RFM* he talked about establishing a form's tautologousness by establishing that it is a theorem of Russell's propositional calculus. We have just seen in section 11 that Wittgenstein's position was that theoremhood in Russell's calculus is established by strict finitistic means. So Wittgenstein and strict finitists would agree that the form (P v~P) is a tautology, and this is proved, not determined by convention.

14. The important question is whether or not they agree on what the tautologousness of (P v~P) shows about the truth of all propositions with this form, and in particular what it shows about the truth of mathematical propositions with this form. A strict finitist would admit that "The number of letters in 'Bismarck' is 8 or the number of letters in 'Bismarck' is not 8" has a tautologous form. He would also admit that the possession of this tautologous form shows that the sentence cannot be used to give information about the number of letters in "Bismarck". He makes this second admission since the rules for assigning truth to disjunctions would assign true to it no matter how many letters "Bismarck" contained. So a strict finitist admits that possession of the (P v~P) form by a disjunction of empirical propositions makes the disjunction an empirical proposition which is always true and hence one which says nothing. What about its status as a mathematical proposition? For a strict finitist a proposition requires a proof to be converted into a mathematical proposition. So for "The number of letters in 'Bismarck' is 8 or is not 8" to be mathematical requires a proof of it. If there could be such a proof that would satisfy a strict finitist, it would be either the proof that its form is a tautology or a strict finitist proof of one of the disjuncts. The proof that its form is tautologous could not be the proof that makes the entire disjunction mathematical because that proof makes mathematical a proposition such as: The truth values assignable to "The number of letters in 'Bismarck' is or is not 8" are true. (Recall the analysis in Chapter III, section 36 on what it is to say that a truth is necessarily true.) (I.e., it changes such a proposition about truth values from a descriptive proposition to an identity. In other words, the proof of the tautologousness of a proposition P makes a metalinguistic truth

about it into a mathematical proposition, but does not make P itself mathematical. A strict finitistic proof of either of the disjuncts will, of course, make that disjunct a mathematical proposition. But making one disjunct mathematical will not make the entire disjunction mathematical. On the contrary, proof of one of the disjuncts makes the entire disjunction meaningless since a strict finitistic proof of a proposition P is based, – causally –, on the fact that its denial is unthinkable. For instance, the strict finitistic proof of "The number of letters in 'Bismarck' is 8" has made us aware that "The number of letters in 'Bismarck' is not 8" is meaningless in the sense that we cannot think of what it would be like for it to be true. So once we prove that "Bismarck" has 8 letters, the disjunction: "The number of letters in 'Bismarck' is 8 or is not 8" is nonsense because the right disjunct is nonsense. So for a strict finitist any proposition of the form (P v∼P) is nonsense if either P or ∼P has been proved. We can look at it this way: A proof of one of the disjuncts turns a tautologous disjunction of empirical propositions into nonsense.

15. There are also sentences of the (P v∼P) form in which neither P nor ∼P has been proved, and neither P nor ∼P has use as an empirical proposition. An example of such a sentence is: The decimal expansion of Pi is a sequence containing 777 or it is not. In this section, let us restrict ourselves to talking about this disjunction concerning the decimal expansion of Pi. Since there is no proof of either disjunct, a strict finitist cannot argue that the whole disjunction is nonsense because he has proved one disjunct to be unthinkable. He cannot claim that the disjunction about Pi is merely a disjunction of empirical propositions such as: Snow is white or it is not. So why wouldn't a strict finitist admit that it is a meaningful mathematical disjunction of two meaningful mathematical statements, which is necessarily true because it has a tautologous form? It certainly sounds mathematical, or, at least, sounds as if it were about something in mathematics. He would not admit this because in the absence of a proof that the decimal expansion of Pi is identical with a sequence containing 777 he has no idea of what it is for the decimal expansion of Pi to be identical with such a sequence. To

be sure he can imagine himself computing Pi, and then imagine himself at some point writing 777. But, although a strict finitistic proof can be in the imagination, it must be surveyable; and this means that one must see the result of an operation being the same as the operating. So imagining oneself writing 777 while computing Pi does not give mathematical meaning to the claim that the decimal expansion of Pi contains 777, unless what one imagined enabled them to see their result of writing 777 as part of the calculating of Pi. If you could see your writing down 777 as part of calculating Pi, you would have calculated Pi to a point where 777 is calculated and thus you would have proved that 777 is in Pi. Arguments to show that we don't understand the affirmative member of the disjunction are sufficient to show that we do not understand the entire disjunction. So a strict finitist would agree with the Intuitionists that this disjunction about Pi is not a meaningful mathematical sentence, because in the absence of proof we do not know what it would be like for there to be 777 in the decimal expansion of Pi.

16. To settle that Wittgenstein's position on the law of the excluded middle is strict finitistic, I must determine his position on what the tautologousness of its propositional form shows us about the mathematical truth of sentences with this form. Next I must determine his position on the meaningfulness of sentences of the form (P v∼P) where one of the disjuncts has been proved. Finally, I must determine his position on the meaningfulness of the disjunction about 777 in the decimal expansion of Pi.

Wittgenstein also held that a proof that the form (P v∼P) is tautologous does not make a sentence of that form into a meaningful mathematical statement. In III-14 he wrote:

> "A tautology (e.g. 'P v∼P') says nothing" is a proposition referring to the language-game in which the proposition P has application. (E.g. "It is raining or it is not raining" tells us nothing about the weather.
>
> Russellian logic says nothing kinds of *propositions* – I don't mean *logical* propositions – and their employment: and yet logic gets its whole sense simply from its presumed application to propositions.

He meant that the metalinguistic claim that a sentence has a tautologous form tells us about the truth of what that sentence could be used to say only in the areas in which the disjuncts already have a use. Thus, unless the disjuncts already have a mathematical use, knowledge of the fact that a disjunction of them is tautologous will not show us that the disjunction is a necessary mathematical truth. He added that a defect of formal logic is that it does not tell us in which areas its propositions are applicable. For instance, there is nothing in logic which tells us that "A sentence with the form (P v∼P) is always true" does or does not apply to mathematics. In IV-12 he made the same point that simply knowing that a sentence has a tautologous form does not suffice for knowing that whatever it is used to assert will always be true; we need also to know how the component propositions are used. He wrote in IV-12:

> In the law of excluded middle we think that we have already got something solid, something that at any rate cannot be called in doubt. Whereas in truth this tautology has just as shaky a sense (if I may put it like that) as the question whether P or ∼ P is the case.

I interpret this passage as saying that it is a contingent matter, – contingent on the use of the disjuncts –, whether or not possession of the (P v∼P) makes a sentence a necessary truth. So Wittgenstein agrees with strict finitism that mere possession of a tautologous form does not make a sentence a necessary truth. In particular, possession of (P v∼P) form does not guarantee that mathematical sounding sentences with this form will be necessarily true or even true.

17. We have seen that a strict finitist holds that sentences with the (P v∼P) form can never be used to assert a meaningful mathematical statement. Did Wittgenstein also hold this? I believe so. I believe that Wittgenstein held that a sentence with the form (P v∼P) cannot be used to state a meaningful mathematical proposition if one of the disjuncts is proved. (You can read much of IV-18 as investigating this problem. Or, as he asked: "isn't the law of excluded middle valid for *these* inside mathematics?") My reason for believing this is that he questioned, as we saw in section 4, REDUCTIO AD ABSURDUM proofs because in them we need to use the denial of a mathematical truth. Also in I-106 to

I-112 he suggested that we cannot believe a false mathematical proposition because it is unthinkable. Certainly, if one disjunct of a sentence with the (P v~P) form were mathematically true, i.e., proved, the other would be a mathematical falsehood and hence unthinkable. But for the whole disjunction to be meaningful that falsehood would have to be thinkable. Wittgenstein also revealed his position that sentences of the (P v~P) form are not meaningful mathematical propositions if one of the disjuncts is proved by making a verbal point in IV-18.

The opposite of "it must not occur" is "it can occur". For a finite segment of the series, however, the opposite of "it must not occur in it" seems to be: "it must occur in it".

In this context propositions about finite segments of a series can be regarded as proved propositions. His point is that negation in a sentence of the form (P v~P) where a disjunct is proved does not have its normal interpretation if the disjunction is to make sense. For instance, after we prove "The number of letters in 'Bismarck' is 8", we can assert it as necessarily true. So after this proof we would have: "The number of letters in 'Bismarck' is 8 or the number of letters in 'Bismarck' need not be 8", if we wanted to apply the principle of excluded middle to this proved proposition. But this disjunction just cited makes no sense because we have no idea of what it is to say that "Bismarck" need not have 8 letters. However, if we write: "The number of letters in 'Bismarck' must be 8 or it must not be 8" we can have a slightly better idea of what we are saying in the right disjunct. We are saying that "Bismarck" cannot have 8 letters. But, of course, this last disjunction cited is not an instance of the law of excluded middle. More generally: a sentence with the (P v~P) form is not an instance of the law of excluded middle if negation does not receive its normal interpretation.

18. Wittgenstein would also agree that the disjunction about 777 occurring in the decimal expansion of Pi is not a meaningful mathematical proposition because in the absence of proof neither disjunct is mathematical. His discussion of the law of the excluded middle runs from IV-9 through IV-28. But since I believe his

position is already brought out in IV-9, my discussion of this issue will center on IV-9. In IV-9 he suggested that a question about the truth of a non-empirical proposition which has no proof, such as "777 occurs in the decimal expansion of Pi", is as meaningless as asking a poet whether or not the hero of his poem has a sister when the poet has not even thought about the matter. Just as the poet can give no answer until he decides whether his hero should have a sister, a mathematician cannot talk about there being an answer until he decides what the answer is to be. Of course, the mathematician's decision cannot be given as freely as the poet's since the mathematician requires proof to make his decision. Wittgenstein put it his way in IV-9:

> The question – I want to say – changes its status, when it becomes decidable. For a connexion is made then, which formerly *was not there.*

This means that sentences such as "777 occurs in the decimal expansion of Pi" change from having no truth value to having one as a result of a proof, and he would say that with no truth value they are not mathematical. As he admitted later in IV-9:

> However, queer it sounds, the further expansion of an irrational number is a further expansion of mathematics.

And expansion of mathematics is augmenting the number of mathematical propositions.

Now Wittgenstein not only agreed with strict finitism that the disjunction about Pi is mathematically meaningless, he also agreed that it is meaningless because in the absence of proof we do not have an idea of what it is for the decimal expansion of Pi to be one with or without 777. In IV-12 he remarked:

> For how do I know what it means to say: the pattern . . . occurs in the expansion? Surely by way of examples – which shew me what it is like for. . . . But these examples do not shew me what it is like for this pattern to occur in the expansion!
>
> Might one not say: If I really had a right to say that these examples tell me what it is like for the pattern to occur in the expansion, then they would also have to shew me what the opposite means.

The last part of the remark means that if a picture of 777 occurring in a sequence of numerals showed us what it is like for there to be 777 in the decimal expansion of Pi, then erasing these three 7s should show us what it is like for there not to be 777 in the decimal expansion of Pi. The embarrassing thing about admitting that we can have a clear idea of what it is like for 777 to be in the decimal expansion of Pi as well as have a clear idea of what it is like for the decimal expansion of Pi not to have 777 in it, is that we admit to knowing what would falsify a necessary truth. One way of making sense of holding that "777 occurs in the decimal expansion of Pi" is necessarily true but that yet we can think of falsifying conditions for it, is to hold also that it describes correctly the infinite expansion already completed. The proposition is necessarily true because the expansion is already completed and is unalterable. But we can imagine how the completed infinite series would look without the 777. Wittgenstein, however, had no use for such platonism. In V-34 he wrote;

> Suppose that people go on and on calculating the expansion of Pi. So God, who knows everything, knows whether they will have reached '777' by the end of the world. But can his *omniscience* decide whether they *would* have reached it after the end of the world? It cannot. I want to say: Even God can determine something mathematical only by mathematics.

So it seems that in regard to the law of excluded middle Wittgenstein can be interpreted as a strict finitist.

19. Let us now see how his objections to Cantor's theorem that the real numbers are non-denumerable and Dedekind's theorem or principle of continuity stem from the strict finitism in his thought. I should warn the reader that I am not mathematically sophisticated. So, if discussion of sophisticated mathematical results by someone not particularly skilled in mathematics disgusts and irritates him, he should turn to section 23.

20. A strict finitist would reject Cantor's theorem because the proof itself is not strict finitistic, and the concept of $2\aleph_0$ which we obtain from the proof has no empirical, – ordinary –, application and hence is not usable for strict finitistic mathematics.

Since it is proved by reductio, the set-theoretic proof of $2^n > n$ is clearly not strict finitistic. The diagonal proof is not strict finitistic despite the fact that in a way it is based on a picture. (If the reader is not familiar with the diagonal proof of the non-denumerability of the real numbers, he should see Appendix II-3 of *RFM* for what a fragment of its proof-pattern is like. Also, most introductory set-theory texts contain it.) For one, the partial list of decimal expansions of real numbers running downwards is not perspicuous as such a list. We cannot tell at a glance that the list is a list of most all of the real numbers between 0 and 1 in their decimal expansion form. We have to be told that the list runs infinitely far down; that is not perspicuous. Similarly none of the finite segments of decimal expansions of the real numbers, running from left to right, are perspicuous as such, since we have to be told that they would be extended infinitely far to the right to uniquely represent real numbers. But the greatest defect, from the strict finitist point of view, in the diagonal proof is that it does not entitle us to identify the operation of making the diagonal numbers different with a result which is different for all the numbers in the list running downward. The reason, from a strict finitistic point of view, that we cannot make such an identification is because we cannot survey this picture to see that a picture of the completed operation is a picture of the desired result. Because the lists are allegedly infinite, there is nothing that we can call a picture of the completed operation of changing each diagonal number.

21. Wittgenstein's discussion of Cantor's diagonal argument is in Appendix II of *RFM*. At first glance, he does not seem to have the strict finitistic objections to the diagonal argument itself. He admitted in Appendix II-2 that the proof gives good grounds for no longer trying to put the irrational numbers in a series. In Appendix II-3, he suggested that the diagonal procedure is a good technique for constructing a real number different from every member of a given series of real numbers. So he did think that the diagonal argument shows something. He thought that it certainly showed a disanalogy between the concepts "cardinal number" and "real number". (By "cardinal number" he meant integer

or natural number.) Nonetheless he did object to the proof. In Appendix II-3 he called it a puffed-up proof.

I submit that his objection to the proof is the interpretation of it as a proof of the existence of transfinite infinite sets. At the end of Appendix II-3 he wrote:

> The difference in kind between the two conceptions is represented, by a shew form of expression, as difference of extension.

What he meant is that we represent the disanalogy between the concepts of cardinal (natural) number and real number as a difference in the number of real numbers and cardinal numbers. I want, however, to show that his objections to Cantor's diagonal argument spring from the strict finitism in his philosophy of mathematics. I can support my interpretation of Wittgenstein as a strict finitist on this issue if I can show that his reasons for rejecting the diagonal proof construed as an existence proof of transfinite sets are strict finitistic. I shall try to show this. We have already noted his claim from IV-41 that concepts which occur in necessary truths must already have a use and meaning in non-necessary, – unproved –, propositions. In V-35, he asked the rhetorical question of what concept can we assert $2^{\aleph_0}$ empirically. For instance, can it be applied to a concept of X number of strokes or can it be applied to some technique as $\aleph_0$ can be applied to the technique of counting for which there is no specified end. I believe that he asks this rhetorical question to suggest that it is useless mathematically because it has no empirical use. (In mathematics we develop rules for the concepts we use outside of mathematics. If a concept has no non-mathematical use, we do not know what rules for its use, developed in mathematics, tell us to do.) Also back in Appendix II-2 Wittgenstein charged that the mistake began when one talked about the cardinal numbers being ordered in a series. (The mistake is asserting the existence of transfinite sets.) I submit that his reason for saying that this is the basic mistake in interpreting the diagonal proof is that it leads us to look at the cardinal (natural) numbers as objects which can be put into a series. Once we have this misleading platonistic picture, it is easy to interpret the difference between the concepts of real and cardinal numbers as there being more real numbers than cardinal numbers. If we would use only number terms which applied to observable pluralities and if, when we talked about all natural numbers, we would talk only of the technique of counting without end, we would not be inclined to think of all the natural numbers as objects which can laid out in a series. We would not be so inclined because our number terms would be adjectives of things which a strict finitist would countenance, viz., observable pluralities such as | | | | or observable techniques, but never unobservable infinite pluralities. Wittgenstein pointed out in Appendix II-17 that it is a mistake to talk about infinities in mathematics if we try to use vague pictures of the infinite, – the enormously big –, to give meaning to mathematical propoistions as the picture  gave meaning to $2 + 3 = 5$ in our strict finitistic proof. In Appendix II-5, 9, 13, and 14, he suggested that "infinite" should be used in mathematics only to classify techniques such as counting with no provision for a last number. If "infinite" is applied to such techniques it gets its meaning from the language of mathematics but does not give meaning to mathematical propositions. But my point is that Wittgenstein wrote as if talking about infinite pluralities in mathematics is objectionable because it leads us to poor strict finitistic mathematics, in other words, to drift into platonism and, worst of all, to determine conceptual connections on the basis of unperspicuous and unsurveyable pictures. So I do not think it rash to conclude that Wittgenstein objected to talk of actual infinities on strict finitistic grounds. Thus we can regard his objections to Cantor's theorem as stemming from the strict finitism in his thought.

22. In *RFM* IV, from 28 through 40, Wittgenstein discussed Dedekind's theorem. In my opinion this discussion is even more difficult to understand than his discussion of Cantor's theorem. In this essay, I shall not examine these Remarks on Dedekind's theorem. I would like to suggest, however, that Wittgenstein's objections to Dedekind's argument in Dedekind's "Continuity and

Irrational Numbers" stems from the strict finitism in Wittgenstein's thought. I suggest that Wittgenstein objections can be read as focusing on one step of Dedekind's proof of the continuity of the real numbers, as Dedekind defined our intuitive concept of continuity. This step is the proof that there are irrational numbers, which proof is a generalization of the proof that the square root of two is irrational. Wittgenstein would not quarrel with the proof of the irrationality of the square root of two. He would quarrel with the standard interpretation of what is proved when we show that the square root of two is not rational. He objected to taking the proof of the irrationality of the square root of two as showing the existence of an irrational number. I believe that he objected to taking the proof of the irrationality of the square root of two as an existence proof because he thought that the reason for taking it as an existence proof is that if we did not we would have to say that a straight line has HOLES in it. I read IV-37 as a claim that we do not have a strict finitistic proof that a straight line has no HOLES.

23. Wittgenstein's remarks on Cantor's and Dedekind's theorems are not the most philosophically impressive sections of *RFM* since they are about aspects of mathematics which one must be mathematically sophisticated to appreciate. Hence, when they are discussed in a mathematically unsophisticated way, as the way in which Wittgenstein discussed them, one has the uneasy feeling that they are not being properly dealt with. However, when we come to his discussion of consistency no special mathematical background is required since he does not talk about the mathematical techniques of consistency proofs. So no one needs to feel unqualified to philosophize along with him here. (His philosophizing against worrying about consistency is primarily in: Appendix I-17, II-78 to II-89, III-56 to III-60, V-8 to V-13, V-21 to V-22, and V-28 to V-32.) At times his position on consistency seems extremely eccentric, but I believe that it can be appreciated if we read him as a strict finitist.

What would a strict finitist say about consistency and is it more or less what Wittgenstein would say? First, to avoid suggesting that anyone advocates mathematicians propounding contradictions, I should note that strict finitists would agree with Wittgenstein in saying that they do not advocate contradictions. Strict finitists would answer a question about using contradictions as Wittgenstein answered it in V-12:

> 'Then are you in favour of contradiction?' Not at all; any more than of soft rulers.

Recall from Chapter II, 29 that for a strict finitist a mathematical truth provides a rule which enables him to determine *a priori* what is the result of the application of an operation to some observable structure. If he had contradictions in his mathematics, he would have the inconvenience of possessing different and conflicting *a priori* determinations of the result of applying the operation just as a ruler which stretches and bends is inconvenient because it gives different measurements of what looks to be the same length. Such results are inconvenient because if we want an unambiguous answer we have to resort to another technique. So a strict finitist's discussion of consistency is academic in that he does not advocate a different use for contradictions than that which they have now. A strict finitist would say with Wittgenstein in II-83:

> It is – I should like to say – for practical, not for theoretical purposes, that the disorder is avoided.

But the disorder is avoided. A strict finitist wants only to change the philosophy of contradictions and their place in nature. He would say with Wittgenstein in II-82:

> My aim is to alter the *attitude* to contradiction and to consistency proofs. (*Not* to shew that this proof shews something unimportant. How *could* that be so?)

(I shall say a word on the parenthetical portion of this passage in section 27). Here the main point to make is that what a strict finitist is trying to do when he talks about consistency is to try to remove what Wittgenstein called in Appendix I-17: "The superstitious fear and awe of mathematicians in the face of contradiction".

24. A strict finitist could say that the belief that there is something mathematically wrong with contradictions as opposed to their practical inconvenience is a superstition. A strict finitist holds that the occurrence of a contradiction does not necessarily indicate that there is any mathematical error in the proofs of the proposition and its denial. I.e, The mathematical reasoning leading to both sides of a contradiction can be perfectly good strict finitistic mathematical reasoning. (Bad strict finitistic reasoning would be reasoning from unperspicuous, unmemorable, or unsurveyable pictures to a proposition which we did not see as being unfalsifiable from the picture.) Such reasoning could be perfectly good strict finitistic reasoning since for strict finitists both a mathematical truth, – proved proposition –, and its denial are empirical propositions before they are made mathematical. There are no *a priori* grounds for holding that a proposition and its denial could not both describe perspicuous, etc. pictures which lead us to make them into mathematical truths. After all, experiments do not have to come out any one way. In I-136, II-78, and V-40 Wittgenstein presented pictures which may lead us to produce mathematics inconsistent with that which we have now. But the occurrence of these pictures does not destroy the pictures we now have for proving the mathematics which we now have. For instance, the picture in I-136 which may lead some of us to assert as necessarily true: $3 + 3 + 3 + 3 + 2 = 10$, would not destroy those pictures which provide strict finitistic proof-patterns for $3 + 3 + 3 + 3 + 2 = 14$. So we could have good mathematics, – no bad proofs –, in which the same numbers had different sums.

I think that Wittgenstein believed that reasoning that leads to inconsistency can be perfectly good mathematical reasoning. He claimed in V-27 that the purpose of the mathematical problem of making a bit of mathematics consistent cannot be to create mathematics where there was none before. So Wittgenstein thought that inconsistent mathematics was mathematics. This together with his belief that a philosopher was to understand and not reform current mathematics suggests that he believed that an inconsistent mathematics could be a perfectly good mathematics. He wrote in V-28:

I mean: if a contradiction were now actually found in arithmetic – that would only prove that an arithmetic with *such* a contradiction in it could render very good service; and it will be better for us to modify our concept of the certainty required than to say that it would really not yet have been a proper arithmetic.

(V-8, V-9, and V-12 contain more remarks that an inconsistent mathematics can be a perfectly good mathematics.) In II-81, Wittgenstein called our attention to the fact that someone could produce by mathematical reasoning a contradiction in a system and be quite proud of this bit of mathematical reasoning. And is this not the way it is? If a man showed that our present arithmetic is inconsistent, he would produce by mathematical reasoning a contradiction.

25. A strict finitist could also say that it is a superstition to believe that contradictions are mathematically wrong because at least one of the contradictory propositions must be false. A strict finitist would call this belief a superstition since it would be based on a belief that one of a pair of contradictory propositions must be false because both cannot correspond to the mathematical facts. What I have just said is not inconsistent with what I wrote in Chapter II, section 35 on strict finitism and mathematical falsehoods. In Chapter II, section 35, I characterized mathematical falsehoods as UNPROVED propositions which are inconsistent with proved ones. In this section we are considering what to say if we have two PROVED propositions which are mutually inconsistent. I think that Wittgenstein's rejection of any platonism in mathematics shows that he would agree that it is mistaken to say that one of a pair of contradictory propositions must be false since they cannot both have mathematical truth conditions. Also in remarks such as:

If the contradiction is so well hidden that no one notices it, why shouldn't we call what we do now proper calculation?

from V-12 and:

The abysses in the calculus are not there if I do not see them!

Is no demon deceiving us at present? Well if he is, it doesn't matter. What the eye doesn't see the heart doesn't grieve over.

from II-78, he showed that he was indifferent to unnoticed or implicit contradictions. If he had thought that one of a pair of contradictory propositions must be false, he would not have been indifferent to implicit contradictions.

26. A strict finitist could also label as superstititous fear the belief that if we have an explicit contradiction everything follows. To be sure one can derive everything from an explicit contradiction. But for a strict finitist, derivations are not mathematical reasoning. For a strict finitist one needs perspicuous, etc. pictures to prove propositions. It is not at all clear that an explicit contradiction provides such a picture for any proposition. I am confident that Wittgenstein was in agreement with strict finitism on this point. In V-9 he suggested that only if we try to do mathematics in a mechanical fashion, viz., solely by derivations from axioms, do we have to worry about concluding everything from an explicit contradiction. He went on to suggest that the idea of doing mathematics mechanically is only a fashion of the times, where I interpret "fashion" as meaning "a foolish fad which will and ought to pass away". As he noted in II-4: Logic is only frills attached to real mathematical reasoning.

So strict finitists and Wittgenstein would agree that inconsistency does not show a mathematical error in that it does not indicate poor mathematical reasoning, it does not show that we have a mathematical falsehood, and it does not yield every proposition as a theorem of mathematics.

27. Wittgenstein and strict finitists would, of course, also agree that consistency proofs are significant bits of mathematics and that on occasion a contradiction can make sense. There is nothing controversial in admitting that consistency proofs can be significant mathematical proofs in that they show that the results of applying the operations called inference rules to the axioms of a system can be only formulae of a certain type.

28. But the claim that contradictions can make sense is controversial. In V14 Wittgenstein wrote:

There is one mistake to avoid: one thinks that a contradiction must be senseless: that is to say, if e.g. we use the signs 'P', '∼' '·' consistently, then 'P·∼P' cannot say anything.

Could a strict finitist agree that contradictions can have a sense? We have seen that a strict finitist would admit that both a proposition P and its denial $\sim$P can be theorems at the same time. But of course to admit this is not to admit that (P$\cdot\sim$P) could be a theorem. For a proposition of the form (P$\cdot\sim$P) to have a strict finitistic proof a proposition of the (P$\cdot\sim$P) form would have to have an empirical use to describe a proof-pattern. It is far more disturbing to admit that a contradiction can be the proper thing to say in an empirical situation than to say that it can have a legitimate mathematical use. This much is clear. If a strict finitist holds that an empirical situation cannot be described by a contradiction, he holds this because he has a strict finitistic proof, most likely from truth tables, that an empirical proposition with the (P$\cdot\sim$P) form cannot be true. In his strict finitistic proof he has taken the empirical truth describing truth tables: "The assignment of truth values to (P$\cdot\sim$P) for all possible values of P is false", and made this empirical truth a necessary truth. But he has to admit that before he made this descriptive truth about truth tables into a necessary truth its denial had a sense. So he has to admit, at least, that if it were not for his proof contradictions could have a sense. And to admit this is to admit very much. A strict finitist always has to be prepared to admit that someone could produce a picture showing what it would be like for his proved proposition to be false. Such a picture would lead him to put his proved proposition back amongst the mere empirical truths, or even on occasion amongst the empirical falsehoods. So a strict finitist has to admit that a new picture could lead him to say that contradictions could be true despite the fact that right now he has not the slightest idea of what such a picture would be. So I think that a strict finitist would agree with Wittgenstein that explicit contradictions could have a sense both in mathematics and in empirical uses although he would say that he cannot now see the sense that they could have. I think that this close agreement on contradictions is strong support for my recommendation that at least much of Wittgenstein's *RFM* can be read as strict finitism.

29. The last issue on which I shall compare Wittgenstein and

strict finitism is on an issue which is discussed more in *PI* than in *RFM*. This issue is: What justifies us in writing a particular number K in the n-th place of a series S(n)? For instance, what justifies us in writing 25 in the 5-th place of the sequence n^2? (In *PI* this is the primary issue discussed in 143 to 156 and in 185 to 241. The four opening Remarks of *RFM* are devoted to this issue and he touched on it in III-9). In I-3 Wittgenstein stated the problem.

How do I know that in working out the series + 2 I must write
"200004, 200006"
and not
"200004, 200008"?

– (The question: "How do I know that this colour is 'red'? is similar.)
"But you surely know for example that you must always write the *same* sequence of numbers in the units 2,4,6,8,0,2,4, etc." – Quite true: the problem must already appear in this sequence, and even in this one: 2,2,2,2, etc. – For how do I know that I am to write "2" after the five hundreth "2"? i.e. that 'the same figure' in that place is "2": And if I know it *in advance*, what use is this knowledge to me later on? I mean: how do I know what to do with this earlier knowledge when the step actually has to be taken?
(If intuition is needed to continue the series + 1, then it is also needed to continue the series + 0.)
"But do you mean to say that the expression '+ 2' leaves you in doubt what you are to do e.g. after 2004?" – no; I answer "2006" without hesitation. But just for that reason it is superfluous to suppose that this was determined earlier on. My having no doubt in face of the question does not mean that it has been answered in advance.

30. I shall support my interpretation of Wittgenstein as a strict finitist by showing that the above passage from I-3 is the natural thing for a strict finitist to say. First, I should like to point out why I believe that Wittgenstein discusses the issue so frequently. He discusses it so frequently because it represents a fundamental break with his position of the *Tractatus* 5.473 that logic takes care of itself. In his later work Wittgenstein shifted to the strict finitistic position that we have to decide what is a truth of logic and mathematics for each and every logical and mathematical truth. So I think that the problem is one which arises from the

position that we must take care of logic. I do not believe that the issue is one of how can we make sure a rule, e.g. + 2, is not interpreted in some unusual way, say as add 4. Wittgenstein admitted he knew what the rule tells him to do; he was not worried about the interpretation of the rule. His worry was about how we know what we have done is in accordance with the rule, given that we understand the rule.

But what about strict finitism and this problem? Once one adopts the strict finitistic position that there are no antecedently existing mathematical facts for our mathematics to correspond with, we have to account for how mathematical propositions can be said to be correct because they do correspond with, or are determined by, a rule. For instance, we would say that writing 25 in the fifth place of the sequence n^2 is the proper answer because that is what we should do according to the rule: square n. So it seems that in mathematics, as it is presently done, there is, in addition to having a proof, another criterion for saying that a bit of mathematics is properly done. This apparently different criterion is that the mathematics, viz., the number computed, does or does not correspond with what should be done according to a rule. Consequently, since a strict finitist holds that provability is the sole criterion for mathematical truth, he will try to explain how the criterion of being correct because it is in accordance with a rule is not a criterion different from that of being provable by strict finitistic means. In particular, a strict finitist wants to avoid the platonistic suggestion that when his calculations are judged by reference to whether or not they accord with a rule, one is checking his calculations by determining whether or not his calculations correspond with a series of numbers already given by the rule in some way. In I-21 Wittgenstein characterized the kind of picture a strict finitist wants to avoid.

In his fundamental law Russell seems to be saying of a proposition: "It already follows – all I still have to do is, to infer it." Thus Frege somewhere says that the straight line which connects any two points is already there before we draw it; and it is the same when we say that the transitions, say in the series + 2, have really already been made before we make them orally or in writing – as it were tracing them.

So if Wittgenstein were a strict finitist it would be appropriate for his reflections on mathematics to begin as *RFM* does begin with an attempt to understand the phrase: "The steps are determined by the formula."

We have now seen that a strict finitist would have the problem of: How do I know that in working out the series + 2, I must write 200004, 200006 rather than 200004, 200008? But the strict finitist is asking how he can know this by strict finitistic means and not have to say that he knows 200004, 200006 is the right sequence because it corresponds with what is determined by the rule +2. Now the question is: Are the other statements made in the passage quoted from I-3 the kind of statement which a strict finitist would make? It shall be easier to understand the remaining statements as strict finitistic statements if I first give the strict finitist solution to he problem.

We know that 200006 follows 200004 in the sequence + 2 because by strict finitistic means we, and almost anyone else who develops the sequence, will when we come to 200004 determine that 200004 + 2 is 200006. There is no question that the rule tells us to add 2 to 200004; that, as I said, is not the problem. But the rule does not give the answer to 200004 + 2. The rule gives only the operation to be applied. At each step we have to determine, – prove –, what is the result of the application of the operation. So a strict finitist can grant that the ordinary language expression: "The steps are determined by the formula" is legitimate in that the formula or rule tells us what operation to use. If we understand our ordinary language, there is no difficulty in knowing which operation the formula or rule tells us to apply. But the expression is misleading insofar as it suggests that the rule can be referred to in some way as a standard for what we have determined to be the result of the operation at each step. Thus the rule "add 2, starting from 0" tells us what operation to apply at the two hundreth and fifty-first step. But this rule cannot be, or in any way provide, a criterion for saying that 504 does not result when we add 2 to 500; only a proof can tell us that.

31. Now let us reconsider the quotation from I-3. Why would a strict finitist say that this problem is like: How do I know that

this color is red? We could say that there is a rule for selecting red objects. This rule runs: The objects similar in color to this paradigm are red. (Each person may have a different group of paradigms, but he does have paradigms.) But this rule gives only an operation; it tells us to compare objects with paradigms to see whether or not they are similar in color. But just as in the case of + 2, knowing what operation is to be performed does not tell us the result of the performance of the operation. So, just as in the case of + 2 we had to go through a proof to determine what results from adding 2, in the case of color judgments we have to use our eyes to determine the results of comparing objects with a paradigm object to tell if they have similar colors. Similarly, the rule: "Add 0 to 2 for each natural number", only gives us the operation "add 0 to 2"; it does not give us the result of any application of this simple operation. We need intuition, *viz.*, strict finitistic proof, to determine the results of any application of + 0 as much as we need it to determine the results of an application of any other operation, *e.g.*, + 1. (But sections 74, 75, and 76 show us that "intuition" is not a good term to use here. The passage from I-3 also suggests that "intuition" is not regarded too favorably.) A strict finitist would also observe that if he were using correspondence with a series already done in some way when he computed a sequence, then he could genuinely doubt that 200006 follows 200004 in the series + 2 because he could always doubt that he is correctly observing the standard or original series. But if he has proved that 200006 follows 200004, there is no doubt. (Indeed the epistemological problem raised in I-3 does not arise from doubts; it arises because there is a case of knowledge about which there is not the slightest doubt. Indubitability is a phenomenon to evoke wonder.)

So I hope that I have shown that a strict finitistic interpretation of the problem and the comments on it are NATURAL. If they are, this supports my contention that it is helpful to read Wittgenstein as strict finitist.

32. What has been accomplished in this Chapter? Certainly, I have shown how several passages from *RFM* can be read if we assume that Wittgenstein was a strict finitist. I hope that this has

shown that it is helpful to read many passages in *RFM* as if Wittgenstein were a strict finitist. Even if Wittgenstein were not simply a strict finitist, if much of his philosophy of mathematics can be construed as strict finitism, this simplifies the problem of finding more exactly what his philosophy was. All that needs to be done is to determine how he went beyond strict finitism. I do not care whether or not I proved that he was nothing but a strict finitist. My purpose has not been to place Wittgenstein into a neat philosophical category and then refute him by refuting that category of philosophic position.

Wittgenstein is what I would call an open as opposed to a closed strict finitist. A closed strict finitist holds that the only good mathematics is strict finitistic mathematics. (We shall see in Wittgenstein's criticism of Intuitionism that he did not want a closed view which would lead him to reforming, or to suggesting mathematical reforms.) An open strict finitist holds that strict finitism is a paradigm of mathematical philosophy. As much as possible we should try to understand mathematics as strict finitists understand the simple bits of mathematics they talk about. We should try to philosophize about other areas of mathematics with as much clarity and with no augmentation of ontology. I would like to close this Chapter with the suggestion that strict finitism plays the same role in *RFM* as behaviorism plays in *PI*. Wittgenstein was more than a strict finitist in *RFM* as he was more than a behaviorist in *PI*. But they were his starting points, and he crept only cautiously beyond them.

IV

WITTGENSTEIN'S OBJECTIONS TO OTHER POSITIONS IN THE PHILOSOPHY OF MATHEMATICS

> If in the midst of life we are in death, so in sanity we are surrounded by madness. (IV-53)

1. Before we assess Wittgenstein's position on the philosophical foundations of mathematics, we shall take a look at his assessment of other positions on the philosophical foundations of mathematics since defects in other positions are some support for his. Also we may improve our understanding of his position, before criticizing it, if we first see what he objected to in other philosophies of mathematics. A man's enemies tell a lot about a man. Logicism is the only position of which he made any sustained critique. *RFM* II from 2 through 56 is devoted primarily to a critique of Logicism. Scattered remarks contain criticisms of Formalism, Intuitionism, and empirical interpretations of mathematical statements as extremely well justified experimental statements or as statements about human behavior. However, I suppose that his remarks against philosophizing against contradiction can be taken as a criticism of Formalism. I shall consider his criticisms of other positions in the following order: Logicism, Formalism, Intuitionism, and empirical interpretations. In this Chapter, I shall assume familiarity with the positions of Logicism, Formalism, and Intuitionism. Max Black's *The Nature of Mathematics* and Stephen Kleene's *Introduction to Metamathematics* contain short accounts of these positions. Stephen Korner's *The Philosophy of Mathematics* contains a longer exposition. And *BP* contains several papers expounding Logicism, Formalism, and Intuitionism.

2. Wittgenstein was not concerned with a technical criticism

of Logicism. For instance, he was not interested in arguing that a logicist must have in his axiomatization of mathematics axioms which are not universally valid formulae. Nor was he interested in arguing, as Gödel did, that not all mathematical truths can be represented as consequences of some set of axioms with which a logicist would begin. As he wrote in V-16:

> It is my task, not to attack Russell's logic from within, but from without.
>
> That is to say; not to attack it mathematically – otherwise I should be doing mathematics – but its position, its office.
>
> My task is, not to talk about (e.g.) Gödel's proof, but to pass it by.

Indeed, Wittgenstein even granted in II-46 and II-53 that every mathematical proof corresponds to a proof in a logicist's axiomatization of mathematics. What, then, are Wittgenstein's objections to Logicism since he was not bothered by the fact that not all axioms required by a logistic axiomatization are universally valid, e.g., an axiom of infinity? Nor was he bothered by the incompleteness and absence of a consistency proof for any logistic axiomatization of all of mathematics. Wittgenstein's objections to Logicism can be classified under two headings. The first is that the logistic axiomatization of mathematics does not provide a foundation for mathematics. (By "logistic axiomatization" I mean an axiom system that a logicist would claim shows that all of mathematics is symbolic logic, as Bertrand Russell said in section 4 of his *Principles of Mathematics*. Or better: An axiom system which a logicist would allege shows that mathematics consists of logical truths obtained by deduction from axioms which are logical truths, all the formulae of the axiom system are written in the notation of symbolic logic, and the formulae contain no undefined non-logical constants. Since any two axiom systems which seem to show that mathematics is symbolic logic would be the same for a logicist, I shall usually use "the logistic axiomatization" rather than "a logistic axiomatization". Of course, I do not mean that a logicist would deny any significant difference between an axiomatic set-theory and the axiomatization in *Principia Mathematica*. There is a difference between starting with properties and starting with classes. I mean that a logicist would not pay any attention to changes in notation and axioms as long as the axioms could give the theorems he wanted. In this essay when I use "the logistic axiomatization" I mean primarily the logistic axiomatization in *Principia Mathematica* because that is what Wittgenstein was talking about. The second is that the mere development of the logistic thesis that mathematics is logic, in works such as *Principia Mathematica,* has corrupted our thinking both in and about mathematics.

3. In II-43 Wittgenstein remarked:

> logic as the foundation of all mathematics does not work and to shew this it is enough that the cogency of logical proof stands and falls with its geometric cogency.

It is obvious from this passage that Wittgenstein claimed that the logistic axiomatization cannot be a foundation for mathematics. It is also clear that he had a reason for this charge. But before we evaluate this, and other, reasons for the charge we must try to understand why he bothered to make the charge since, as he claimed in V-13, mathematics needs no foundations. Indeed this charge that the logistic axiomatization is not a foundation for mathematics seems at odds with the passage we recently cited from V-16 where he claimed that he was not interested in attacking the axiomatization as a mathematical instrument, viz., attacking it from within.

We have already noted in Chapter I that "foundations of mathematics" is ambiguous. It can mean either mathematical foundations of mathematics or philosophic foundations of mathematics. A logistic axiomatization provides a mathematical foundation for mathematics if developing it results in clear concepts of such mathematical terms as "point", "limit", and "number"; if developing it uncovers fundamental principles of mathematics; if developing it leads to greater rigor; if the developed axiom system is complete with respect to the mathematical truths; and if the axiom system is consistent. For an axiomatization of mathematics to provide a mathematical foundation for mathematics, it must be mathematics; but, of course, mathematics done better in regard to rigor and clarity than it is actually done. If the axiomatization were not mathematics, establishing its consistency would not show the consistency of our actual mathematical reasoning; nor would the clarity of its concepts and the rigor of its proofs be any improvement in mathematics. A logistic axiomatization provides a philosophic foundation for mathematics if we can regard it as sufficiently like mathematics and if, when we so regard it, we can give satisfactory answers to the three groups of questions on the philosophic foundations of mathematics, which were raised in Chapter I, section 1. As I discussed philosophic foundations of mathematics in section 31 of the same Chapter, it is not obvious that a logistic axiomatization must be mathematics to provide a philosophic foundation for mathematics; it need only be sufficiently like mathematics. Recall that in Chapter I, section 27 we observed that Wittgenstein held only that mathematics needs no mathematical foundations. So his remark in II-43 that logic cannot provide a foundation for mathematics is inconsistent with the rest of his work only if he meant "mathematical foundations of mathematics".

Incidentally, we can note here that if we accept that Wittgenstein was a strict finitist we can understand why he was indifferent to mathematical foundations. (From here on, I shall write "mathematical foundations" in place of "mathematical foundations of mathematics" and "philosophic foundations" in place of "philosophic foundations of mathematics".) Consistency, as we have seen, is obnoxious only for practical purposes and for practical purposes we can ignore contradictions. Of course, since a strict finitist does not prove theorems by deriving conclusions from premisses, a strict finitist would be unconcerned with logical rigor. Definitions of mathematical concepts are uninteresting since a finitist is interested in how concepts in empirical use are connected with other concepts in empirical use; and proofs, not definitions, give him these connections. Since a strict finitist makes mathematics as he proceeds, he would not want mathematics to be complete, viz., completely finished, if such a notion would even make sense to him.

In my opinion, much of *RFM*-II is devoted to showing that the logistic axiomatization of *Principia Mathematica* cannot provide a mathematical foundation because this axiomatization is not and cannot be mathematics. So I do think that Wittgenstein's charge in II-43 that the logistic axiomatization is not a foundation for mathematics is inconsistent with claims such as those in V-13 that he is not interested in mathematical foundations. Of course, this inconsistency is trivial for Wittgenstein's philosophy of mathematics. The interesting questions for interpreting *RFM* are as follows. What are his arguments that the logistic axiomatization is not mathematics? Are these arguments inconsistent with his claim that he wanted to show by non-mathematical means that the logistic axiomatization does not and cannot perform its office? And of most importance: Do these arguments show that the logistic axiomatization of mathematics is so much different from actual mathematics that discussion of it cannot provide philosophic foundations?

4. Wittgenstein's conclusion that the logistic axiomatization is not mathematics rests on two claims about the logistic axiomatization. The first claim is that when we work with the logistic axiomatization, the mathematics that we actually do is only a small part of mathematics and that small portion is in the metalanguage of the logistic axiomatization. The second claim is that we could not actually do any mathematics within the axiomatization itself. The interest of these claims that the logistic axiomatization is not mathematics is that they establish that the logistic axiomatization cannot perform the office of providing mathematical foundations. They show that the logistic axiomatization cannot provide mathematical foundations since, as we shall see, only mathematics can provide such foundations. As we shall also see, these two claims provide the basis for what can be interpreted as Wittgenstein's argument that the logistic axiomatization cannot provide philosophic foundations.

The charge, which I just quoted from II-43, that the cogency of logical proof depends upon its geometrical cogency is part of an argument to support the first claim. As I showed in our discussion of the law of excluded middle in Chapter III, section 10, the argument is that the mathematics we do when we deduce formula F from the axioms is to establish the mathematical truth, – metamathematical truth –, that the sequence of formulae which we construct is a formal proof of the formula F. As a matter of fact, Wittgenstein also believed that the logistic axiomatization was not even very good for this metamathematics. When the formal proofs become extremely long we cannot establish with certainty that they are formal proofs of the end formula. Wittgenstein pointed this out in II-2, II-7, and II-14. He pointed out that in very long formal proofs we have only empirical certainty that the sequence which we have constructed is a proof of the end formula. This is not merely to say that we have only empirical certainty about what is written down. It is to say that we cannot be certain that what we are aware of is a proper formal proof, despite the fact that we do not worry about seeing the formulae in the proof sequence properly. He added that if we abbreviate the techniques of proof in a formal axiom system so that, for instance we can eliminate a hundred steps, then we may have a proof that a certain formula is formally provable by means of the abbreviated techniques although we could get no proof that the formula is provable by means of the unabbreviated techniques. As Wittgenstein also wrote in II-43:

> logical proof, e.g. of the Russellian kind is cogent only so long as it also possess geometrical cogency. And an abbreviation of such a logical proof may have this cogency and so be a proof, when the Russellian construction, completely carried out, is not.

The significance of this second charge from II-43 that formal proofs carried out by unabbreviated rules of proof are not even good proof-patterns for metamathematical propositions that the end formula of a sequence is formally proved by the sequence is that it is a charge that completely rigorous formal proofs, – ones with all the steps –, have no mathematical office. Or at best they have an insignificant office. The rigor of formal proofs is only a pretense of rigor. The formal proofs, as shall be argued, are not mathematical proofs and as we have just seen the formal proofs are not even part of good metamathematical proofs. So if they have any rigor, it is not the rigor of anything mathematical. (A metamathematical proposition is a mathematical proposition in a metalanguage.)

5. In my opinion, Wittgenstein was correct in holding that the cogency of logical proofs, i.e., proofs in a formalized axiom system or uninterpreted calculus, are dependent upon their geometric cogency. In an uninterpreted calculus a proof sequence, if it shows anything, shows that the last formula in the sequence is a theorem of the calculus. A formula's being a theorem of a calculus is not what we would regard as a contingent matter. Whether or not a certain formula can be derived from the axioms is as fixed as whether or not REVO can be obtained as the reverse of OVER. So showing that a formula is a theorem of a calculus is a topic of genuine mathematical concern even if not of great importance. We certainly would not be satisfied with mere empirical certainty that a formula is a theorem. But it is a fact that if a proof sequence in an uninterpreted calculus is so long, viz., lacks geometrical cogency, that I cannot be certain that it is a proof sequence, I cannot assert as a mathematical truth that the end formula of the sequence is a theorem of the calculus. However, on occasion I can abbreviate the rules of proof and be certain that a sequence constructed according to the abbreviated rules is a proof according to the abbreviated rules.

6. Wittgenstein was also right in holding that logic cannot be a mathematical foundation because the cogency of logical proof depends upon geometric cogency. (Recall that the gist of this charge that logical cogency depends upon geometric cogency is that the mathematical office of the logistic axiomatization is only to provide proof-patterns for metamathematical propositions of the form: Formula F is a theorem of axiom system A.) He would be right on this point even if every logical proof provided complete geometric cogency that its end formula is a theorem. Even if every formal proof provided geometric cogency that it was a proper proof sequence, this would show only that the axiom system was a very good device for providing proof-patterns for metamathematical propositions about what is provable in the axiom system. Now providing proof-patterns for the proofs of such metamathematical propositions could not be providing mathematical foundations. These metamathematical propositions

constitute only a very small part of actual mathematics. Proving them well is not going to provide much support for the rest of actual mathematics.

It should be noted that although much of Wittgenstein's criticism of Logicism in II-2 to II-56 can be read as a characterization of how a strict finitist would do the metamathematics of *Principia Mathematica,* and a complaint that *Principia Mathematica* is not a good system for doing metamathematics in a strict finitistic way, the charge that logic is not the mathematical foundation of mathematics because logical cogency depends upon geometric cogency is not based on a premiss that metamathematics should be done in a strict finitistic way. It is based on a premiss that proving propositions of the form: "Formula F is a theorem of axiom system A", is insignificant in providing mathematical foundations.

7. However, someone may make a case that proving propositions of the form: Formula F is a theorem of axiom system A, is significant in providing mathematical foundations, especially if that axiom system is that in *Principia Mathematica.* (Hereafter, I shall call the axiom system of *Principia Mathematica, PM.*) They would point out that when *PM* is given its normal interpretation, every mathematical proposition can be correlated with a distinct interpreted formula of *PM.* I should point out that here "formula of *PM*" refers to well-formed formulae and not just theorems of *PM.* Also I should point out that "normal interpretation of *PM*" means only interpreting the logical signs, *viz.,* giving a reading to the quantifiers, propositional connectives, and the way of representing that something has a property or that somethings stand in a relation. So no interpreted formula of *PM* would naturally read as a mathematical proposition. At least no interpreted formula of *PM* would sound like the kind of proposition actually used in mathematics. Wittgenstein himself granted in II-2, II-46, and II-53 that there could be such a correspondence between propositions of actual mathematics and interpreted formulae of *PM.* Indeed it seems that he not only admitted that every mathematical proposition can be re-expressed in the language of *PM* but, contrary to Gödel's theorem, every proof in actual mathe-

matics can be rewritten as a formal proof in *PM*. But let us return to my hypothetical defender of the logistic axiomatization. This defender of Logicism would go on to point out that the mere fact that mathematical propositions can be re-expressed as interpreted formulae of *PM* solves one problem of mathematical foundations. This is the problem of providing clear definitions of mathematical concepts. The very task of re-expressing propositions of actual mathematics requires that all mathematical concepts be defined in terms of logical ones. So Logicists have certainly given clear definitions of mathematical concepts. Also once we have this correspondence between propositions of actual mathematics and interpreted formulae of *PM* we can identify the axioms and inference rules of *PM* as corresponding with, viz., being the re-expression of the axioms, possibly implicit, and inference rules used in actual mathematics. Thus, defenders of *PM* as providing mathematical foundations would allege that to know that formula F is a theorem of interpreted *PM* provides knowledge that the ordinary mathematical language correlate of formula F is rigorously provable from the axioms and inference rules of actual mathematical practice. Hence these defenders of the mathematical office of *PM* could also go on to claim that a consistency proof of *PM* would show the consistency of a large portion of actual mathematics.

8. However, someone who gives the above type of defense of *PM* as providing mathematical foundations is assuming that mathematics re-expressed as interpreted formulae of *PM* is still mathematics. If the interpreted formula F which re-expresses a mathematical proposition M were not merely another version of M, then knowing that F is formally provable from the formal axioms would not establish that M can be rigorously proved from the informal versions of the formal axioms. Hence, we can see how the second claim presented at the beginning of section 4 of this chapter supports the conclusion that the logistic axiomatization cannot provide mathematical foundations for mathematics. This is so because, in effect, this second claim is a claim that the interpreted formulae of *PM* which re-express mathematical propositions are not a version of the mathematical propositions which

they re-express. Let us look at Wittgenstein's arguments for this second claim, i.e., the claim that we cannot do actual mathematics within *PM*. This will show why, and what it means to say that, the logistic axiomatization is not another versioo of mathematics.

9. Wittgenstein's claim that we cannot do much, if any, mathematics withing the logistic axiomatization is supported by the following three points. The first point is: If we originally possessed the logistic axiomatization but had not yet done any of what we now call mathematics, all of what we now call mathematics would remain to be done. In II-47 Wittgenstein wrote:

> Can what I want to say be put like *this*: "If we had learnt from the beginning to do all mathematics in Russell's system, the differential calculus, for example, would not have been invented just by our having Russell's calculus

In II-46 he also remarked:

> Now surely one could simply say: if a man had invented calculating in the decimal system – that would have been a mathematical invention! – Even if he had already got Russell's PRINCIPIA MATHEMATICA.

(There are other comments like these in II-46 and II-56.)

In my opinion, Wittgenstein was absolutely correct in maintaining that we would not even have begun to develop what we now call mathematics if we had only deductions within *PM*. Of course, if we had also some metalinguistic truths about *PM* we would have a small fragment of what we now call mathematics. But this metamathematics is not mathematics within *PM* even if it can be re-expressed in *PM*. The way to convince oneself of this point is to ask oneself how much mathematics one would know if all that he were taught of *mathematics* was deductions in interpreted *PM*, where "interpreted" means only that the logical signs have been given a sense. Knowledge of such deductions would not even give one knowledge that $2 + 2 = 4$.

10. The second point is: Although once branches of actual mathematics, such as computations in the decimal system and

the differential calculus, have been developed we can find the correlates of this mathematics in *PM*, the formal proofs in *PM* are not the proofs that established these results in actual mathematics. In other words, the second point is what Wittgenstein suggested in II-4. In II-4 he suggested that the logical calculus, i.e., *PM*, is a frill tacked on to actual mathematics. The logical calculus is a frill because the logical calculus is parasitic on actual mathematics and does not improve it. Only after we have established truths in actual mathematics can we reconstruct their proofs and restate them in the logical calculus. Wittgenstein developed this point in II-46. In II-46 Wittgenstein admitted that the work in *PM* corresponding to a branch of mathematics could be called a new aspect of that branch. "Aspect" is a vague term here. I do not think that to say that proposition P is proposition Q under a new aspect is to make the strong point that P and Q say the same thing. However, in this context, it is not important if we give "aspect" this strong interpretation. After this unclear admission about *PM* being as aspect of actual mathematics, he considered the following suggestion.

"But in that case the truth of true mathematical propositions can still be proved from those general foundations."

(Of course, the general foundations are the proofs in *PM*.)

His reply to this suggestion was:

It seems to me there is a snag here. When do we say that a mathematical proposition is true?

The snag is that we say that a mathematical proposition is true when we have proved it. Thus, to prove true propositions in *PM* is redundant since the fact that they are true shows that they have already been proved.

I think that Wittgenstein was correct on this second point that work in *PM* is redundant for the actual development of mathematics. To convince ourselves of the redundancy of the logistic axiomatization of mathematics, let us consider how we would answer the following two questions in this passage from II-53.

"But surely we can always go back to the primitive logical method!" "Well assuming that we can – how is it that we don't have to? Or are we hasty, reckless, if we do not?

Who would want to maintain that if mathematicians were really doing their jobs, they would be doing nothing but deductions in *PM*? And if we would not maintain this, is it not because mathematicians are already doing their job properly?

11. The third point is: The proofs in a logistic axiomatization cannot replace those in actual mathematics because we still need to use ordinary mathematics to determine whether or not the logistic proofs give the proper result. Wittgenstein raised such a point in II-17 when he asked:

Let me ask the following question: Is it not conceivable that the 1-1 correlation could not be trustworthily carried out in the Russellian proof, that when, *for example*, we try to use it for adding, we regularly get a result contradicting the usual one, and that we blame this on fatigue, which makes us leave out certain steps unawares? And might we not then say: – if only we didn't get tired we should get the same result – ? Because *logic* demands it? Does it demand it, then? Aren't we here rectifying logic be means of another calculus?

In this passage from II-17 he was saying that when we try to add large sums by means of doing a deduction in *PM*, we may easily get confused by the complicated formulae. Thus, we use ordinary arithmetic techniques to decide whether or not an addition done in *PM* is correctly done. So we cannot say logic, – deductions in *PM* –, really justifies what we do in addition. In II-4, II-13, and II-14 he also makes this point that if we deduce results in *PM*, or some other logistic axiomatization, inconsistent with results obtained by ordinary mathematical techniques, we say that the results obtained in the formal system are wrong.

Of course, a user of the formal axiom system would not say that the formal axiom system yielded mathematical falsehoods. He would say that he miscalculated in the formal system whenever his formal calculations conflicted with the results of ordinary calculations. In regard to the opening quotation of Chapter I, we can see here that there is a use of "I must have miscalculated" which reveals something of importance about the mathematical foundations of mathematics. In the context in which someone rejects a formal deduction with "I must have miscalculated" because the formal deduction conflicts with an ordinary calculation, his use of "I must have miscalculated" reveals that the logistic axiomatization cannot be a mathematical foundation since actual mathematics has a legislative control over formal mathematics.

To me, this third point seems correct. Who would accept that $10^{10} + 1 = 10^{10}$ simply because to the best of our knowledge the *PM* version of $10^{10} + 1 = 10^{10}$ was provable? It does not help *PM* to say that the proofs in *PM* are still logically prior even if they are not psychologically prior. If we use the PSYCHOLOGICALLY prior proofs, – proofs actually used in mathematics –, as a standard for correctness of formal proofs, the psychologically prior proofs are logically prior to the formal proofs.

12. The above three points establish that working within a logistic axiomatization is not doing any mathematics. The first and third points show that such work is not sufficient for doing actual mathematics while the second point shows that work in a logistic axiomatization is not necessary for doing any actual mathematics. If working in a logistic axiomatization is neither necessary nor sufficient for doing actual mathematics it follows that not much if any actual mathematics can be done in the logistic axiomatization. Actually the stronger conclusion that no mathematics can be done in the logistic axiomatization is yielded by the three points. But I use the weaker "not much if any" because Wittgenstein granted in II-4 that very small computations such as $2 + 3$ can be done within *PM*. Still, I think that what he really meant was that a *PM* formula expressing $2 + 3 = 5$ could be used as a strict finitistic proof-pattern for $2 + 3 = 5$, and the proved proposition would be in the metalanguage. So I submit that Wittgenstein held, in effect, that no mathematics could be done within a logistic axiomatization of mathematics. And it seems to me that this is true. However, as I pointed out in Chapter I, section 7 we cannot prevent people in the future calling work within *PM* mathematics. If in the future work within *PM* is called

mathematics and treated as mathematics, then a philosophy of mathematics will have to cover it.

13. We have seen that Wittgenstein has shown that a logistic axiomatization cannot have the office of providing mathematical foundations because it is not mathematics. Our next question is whether or not he has done this by attacking it from without or within. Recall from section 2 of this chapter the quotation from V-16 in which he claimed that he wanted to attack *PM* from without, i.e., non-mathematically. In my opinion the attack was from without. He did not argue mathematically that a logistic axiomatization such as *PM* is mathematically inadequate to provide mathematical foundations. For instance, he did not bother to try to prove the inconsistency or incompleteness of *PM*. He showed in a non-mathematical way that *PM* is inadequate to be mathematics. What was this non-mathematical way? Wittgenstein's arguments that *PM* is not mathematics are of the type that they should be for a philosopher who held that a philosopher should describe and not explain. (Mathematics is not describing.) He described, or more accurately, he recommended that we describe *PM* and actual mathematics, – get clear about what they are –, and the descriptions will remind us that they cannot be the same. If it is a philosophic task to show that *PM* cannot be a mathematical foundation for mathematics, we can again appreciate Wittgenstein's claim in *PI*-127 that a philosopher should assemble reminders for a particular purpose. The descriptions we give ourselves of *PM* and actual mathematics remind us that they are not the same, and hence that *PM* is not a mathematical foundation.

14. Book II of *RFM* was written around 1940. In the March 1958 issue of *The Review of Metaphysics,* Alan Ross Anderson published his "Mathematics and the Language Game", which is an unfavorable review of *RFM*. (It is reprinted on pp. 481-490 of *BP*.) Anderson charges that Wittgenstein's remarks on logicism are out of data since for the past twenty-five years no one of any ability has seriously maintained that mathematics can be reduced to logic. Anderson points out on pp. 484-485 of *BP* that the contemporary position is " 'mathematical logic' is logic studied

mathematically, and that using mathematical techniques to study the syntax of artificial languages does not somehow count as 'reducing' mathematics to logic." Of course, Wittgenstein was not quite as many years behind the times as Anderson charges. But even if Wittgenstein's attack on the position that mathematics can be reduced to logic were an attack on a view no longer widely held, I think that his attack has value. (My personal opinion is that the view that mathematics can be reduced to logic is very widely held although maybe it is not held by the best logicians.) Wittgenstein's attack is valuable because it shows that we do not need mathematical results to conclude that mathematics is not and cannot be logic, viz., a logistic axiomatization. At least at the present time, we cannot identify logic with mathematics. But despite the value of establishing by non-mathematical means that mathematics cannot be reduced to logic, and hence that a logistic axiomatization cannot be a mathematical foundation, the entire issue of finding mathematical foundations may seem of little importance. May it not still be the case that a logistic axiomatization, such as *PM*, can provide a philosophic foundation for mathematics? For instance, could not the fact that no descriptive constants need be used in *PM* be grounds for the ontological claim that mathematical propositions are not truths about some objects? Or could not the fact that almost all of the re-expressions of actual mathematical propositions are universally valid formulae be grounds for the epistemological claim that mathematical necessity is logical necessity? Indeed, it is not immediately obvious that *PM*'s failure to provide a mathematical foundation shows that it cannot provide a philosophic foundation. Of course, *PM* would not be the philosophy of mathematics; our answers to the three groups of questions from section 1 answered by reference to *PM* would be the philosophy of mathematics. However, we can find in *RFM* remarks that establish that a logistic axiomatization cannot provide philosophic foundations either.

15. In *RFM* the remarks that establish that a logistic axiomatization is not mathematics are those which establish that a logistic axiomatization does not provide a philosophic foundation for mathematics. It would be misleading for me to suggest that

Wittgenstein made the distinction between mathematical and philosophic foundations. I submit that when he claimed in V-16 that he was going to show by external means, i.e., non-mathematical means, that a logistic axiomatization could not fulfill its office he meant that he was going to show that such an axiomatization could provide neither a mathematical nor a philosophic foundation for mathematics. So now we have to look at his non-mathematical arguments that *PM* is not mathematics as showing that *PM* is not sufficiently like mathematics so that it can provide philosophic foundations for mathematics. This phrase "sufficiently like" is annoyingly vague; but I use it so that we will not restrict a philosopher of mathematics to talking only about the sentences actually used in mathematics expressed exactly as they are used. Some alterations may help us appreciate philosophic points. But to discuss Wittgenstein's arguments that *PM* cannot provide philosophic foundations for current mathematics because *PM* is not sufficiently like current mathematics, I must at least give a necessary condition for "sufficiently like". If we re-express an actual mathematical proposition M as a proposition M-1 where M-1 is not a proposition actually used in mathematics, M-1 is sufficiently like M only if the mathematics done with M could have been done if we would have originally started out with M-1. For instance, if M is $2 + 3 = 5$, candidates for a suitable M-1 are the strict finitistic "The counting together of two and three is counting five" and the *PM* language re-expression of $2 + 3 = 5$. Certainly this condition for applying "sufficiently like" is necessary. You could not find out anything about mathematics by talking about forms of expression which could not even be used to do mathematics. Of course, once we recognize this necessary condition for applying "sufficiently like", we use Wittgenstein's arguments that logic is not mathematics to conclude that logic is not enough like mathematics to provide a philosophic foundation for mathematics. This is especially true of the arguments from section 9 of this chapter, where we noted that the *PM* correlates of theorems in the differential calculus could not be used to do calculus.

We should note that to establish the philosophic irrelevance of logic, e.g., *PM*, it was very important to use non-mathematical arguments. If Wittgenstein had shown only that *PM* was mathematically inadequate, he would not have shown that *PM* could not be used to do a portion of mathematics. For instance, if he had shown that it was incomplete or inconsistent, he would not have shown that *PM* could not be used to do some mathematics, or when slightly altered to make it consistent, to do some or all mathematics. If *PM* could be used to do a portion of mathematics, it could still have the office of providing a philosophic foundation for that portion. As we shall see in the next chapter, Wittgenstein believed that providing a philosophy of a portion of mathematics is significant. So before he would dismiss *PM*, he would want to be sure that *PM* is philosophically irrelevant for anything that we now call mathematics.

16. In *RFM* we can find two other kinds of reasons for holding that a logistic axiomatization should not be used for developing a philosophy of mathematics. The first is: The belief that mathematics can be properly expressed as logical truths in a logistic axiomatization is based on an incorrect metaphysics. The second is: If we believe that mathematics can be properly expressed as logical truths in a logistic axiomatization, we get a distorted idea of what mathematics actually is.

17. In IV-48 Wittgenstein wrote:

> 'Mathematical logic' has completely deformed the thinking of mathematicians and of philosophers, by setting up a superficial interpretation of the forms of our everyday language as an analysis of the structure of facts. Of course in this it has only continued to build on the Aristotelian logic.

This passage is relevant to the following kind of defense of a logistic axiomatization as providing philosophic foundations. A philosopher may hold the view that to get the correct ontology, viz., to determine what there must be for there to be the world that we experience, we must rewrite the sentences of our ordinary language in their logical form. A sentence S is written in logical form if it is written as LS, LS has the same truth conditions as S, but as long as S can be either true or false it is impossible that any of the referring terms of LS fail to have referents although the referring terms of S may fail to have referents. For instance, if S is in Russell's famous "The present King of France is bald", his analysis of it as "There is something X is bald and a King of France and anything which is a bald King of France is X" has the logical form of S. Or at least it is closer to the logical form of S. (We cannot definitely say that Russell's analysis gives the logical form because we cannot be sure that the predicate "King of France" has a single quality as a referent.)

A re-expression of the sentences of our ordinary language in logical form can be called an Ideal Language. Once we have an Ideal Language, a correct ontology can be developed simply by listing what kinds of things must be referred to in order to use the Ideal Language. (Since I am here presenting a criticism of philosophizing by means of an Ideal Language, it is not necessary to specify what devices in a sentence of an Ideal Language would be used to refer if the Ideal Language were used.) A philosopher who philosophizes by means of an Ideal Language may be inclined to say that the Ideal Language shows him the structure of facts since the Ideal Language shows him what there must be for there to be the kinds of facts he experiences; it shows him the simples composing the facts and how these simples are connected to form facts. (Crudely put: An Ideal Language philosopher believes that God made the world in the image and likeness of the Predicate Calculus.)

An Ideal Language philosopher may claim that he does not absolutely need the notation of symbolic logic to state sentences of his Ideal Language. However, the multiplicity of ways of making assertions in natural languages makes it seem very unlikely that any single natural language or way of speaking within a natural language would show the structure of facts, if there is such a structure to be shown. I do not think it is unjust to accuse the rigid notations of mathematical logic, even Aristotelian logic, of giving birth to the idea of an Ideal Language. It may be unjust to accuse logic of deforming the thinking of mathematicians and philosophers since the notation of logic does not represent itself as showing the structure of facts. By itself, mathematical logic is simply an extension of Aristotelian logic and Aristotelian logic is merely a tool for helping us assess deductive validity.

In his *Tractatus* Wittgenstein was an Ideal Language philosopher. In his *Philosophical Investigations,* especially in 37 to 65, he rejected the metaphysics underlying Ideal Language philosophy. This underlying metaphysics is the belief that there are objects which must exist if there are to be the kinds of facts we experience. These objects which must exist could be called basic objects or simples. This belief in simples implies that there is one correct description of the world. The one correct description would, of course, be a description of the order and connection of the simples, or any description whose logical form was a correct description of the order and connection of the simples. (As we shall see in the last chapter, strict finitism is inconsistent with the metaphysics underlying Ideal Language philosophy.)

But what has the rejection of Ideal Language philosophy to do with rejection of *PM* as providing philosophic foundations of mathematics? Philosophizing by talking about a logistic axiomatization, or at least mathematics re-expressed as logical formulae, would be how an Ideal Language philosopher philosophizes about mathematics. An Ideal Language philosopher would try to define as many referring terms as he could since if a term used to refer can be defined that means that a complex of conditions must be present for its referent to be present. In other words, the referent of a definable referring term is not a simple. I think that Frege and Russell have shown that if we define all mathematical terms which seem to refer we will re-express all mathematical truths in formulae which for the most part are logical truths, and in which no referring term need occur unbound by a quantifier. So *PM* is how mathematics would look in an Ideal Language. Hence rejection of Ideal Language philosophy is a rejection of one of the best reasons for philosophizing about mathematics by means of talking about a logistic axiomatization.

18. We shall now turn to an investigation of the charge that believing that mathematical propositions can be properly expressed as logical truths in a logistic axiomatization gives a distorted view of mathematics. In IV-24 Wittgenstein wrote:

The harmful thing about logical technique is that it makes us forget the special mathematical technique. Whereas logical technique is only an auxiliary technique in mathematics. For example it sets up certain connexions between different techniques.

It is almost as if one tried to say that cabinet-making consisted in glueing.

For instance, a mathematician may establish several Lemmas by his mathematical techniques and then use logic to deduce his Theorem from his Lemmas. It would be odd to say that all his reasoning was like the deduction of the Theorem from the Lemmas. I do believe that Wittgenstein was correct in claiming that accepting *PM* as providing a philosophic foundation for mathematics can lead to the distorted epistemological view that if mathematicians did their job properly they would simply derive conclusions from premisses. Even if one is a non-mathematician, as I am, a brief look at what good mathematicians do in their lectures and papers makes it seem preposterous that anyone could believe that the real justification of a theorem lies in the deduction of a logistic re-expression of it from some axioms. Yet, I fear that such a belief is still widely held because it is still believed that *PM* provides a philosophic foundation for mathematics.

Wittgenstein also claimed that the belief that mathematics can be expressed as logical truths in a logistic axiomatization gives us a distorted idea of the conditions which make a proposition a meaningul mathematical proposition. In IV-46 he wrote:

'The curse of the invasion of mathematics by mathematical logic is that now any proposition can be represented in a mathematical symbolism, and this makes us feel obliged to understand it. Although of course this method of writing is nothing but the translation of vague ordinary prose

The kind of proposition Wittgenstein was particularly concerned about was assertions of the existence of a number which we do not know how to compute. An assertion that 777 occurs in the decimal expansion of Pi is an example of the kind of proposition that Wittgenstein had in mind. Let us call "777 occurs in the decimal expansion of Pi" B. In the notation of mathematical logic

B could be represented by a well formed formula. Let us call the well formed formula representing B: FB. Since FB is a well formed formula which can be interpreted we believe that it can be understood when interpreted. But what should we understand when we read the interpreted FB? The formula representing B, when read in ordinary language, may not sound at all like B. So at first glance, it may seem inaccurate to claim that the formulae of mathematical logic only translate our vague ordinary prose. However, FB will begin with an existential quantifier, or its equivalent, and this quantifier only translates the vague ordinary "there is" to "exists". Consequently, when we read FB it will tell us in a complicated way that there is a 777 in the decimal expansion of Pi. But now because B is expressed so impressively as FB we feel that it must say something; it looks just like all the other formulae. The only thing we can think of having FB say is that 777 exists in the decimal expansion of Pi, and because "exists" has its ordinary sense we are saying that 777 is somewhere in the decimal expansion of Pi. But as we have seen in the discussion of the law of excluded middle in Chapter III, section 15, we do not know what it is like for there to be 777 in the decimal expansion of Pi unless we know how to compute 777 as part of the decimal expansion of Pi. If we do not know what it would be like for B and FB to be true, we do not understand them. A *reductio ad absurdum* proof of FB would be a perfectly good formal proof of FB. Still it would not give us understanding of FB since it would not show us what it is like for it to be true. But if we had a formal proof of FB we would think we had something that had to make sense.

I think that the preceding discussion of B and FB show the justice of the following complaint from II-25.

A psychological disadvantage of proofs that construct *propositions* is that they easily make us forget that the *sense* of the result is not to be read off from this by itself, but from the *proof*. In this respect the intrusion of the Russellian symbolism into the proofs has done a great deal of harm.

Of course, he meant that proofs in the Russellian symbolism do not give a sense to the proved proposition and thus the proofs do not provide understanding of the propostion. And if no one thought that *PM* had a philosophic office or mathematical office there would be no intrusion of the Russellian symbolism into mathematics.

19. I have now completed my examination of Wittgenstein's devastating critique of Logicism. I submit that he has reminded us, by getting us to take a closer look at what is done in actual mathematics and what is done in a logistic axiomatization, that a logistic axiomatization is so unlike actual mathematics that not only can it not provide mathematical and philosophic foundations for mathematics but also that the serious attempt to reduce mathematics to logic has produced a distorted view of mathematics. We shall now turn to his critique of Formalism and Intuitionism.

20. *RFM* does not contain any long critique of Formalism as it does of Logicism. Although, of course, as we noted in section 1 of this chapter, the recurring attacks on the position that mathematics needs a consistency proof to be certified as being proper mathematics can certainly be regarded as criticisms of the goal of Formalism. In his paper "On the Infinite", translated and printed on pp. 134 to 151 of *BP*, David Hilbert stated the goal of Formalism as:

> The goal of my theory is to establish once and for all the certitude of mathematical methods.

This certitude would be established by showing the consistency of the principles used in mathematics. However, Formalists planned to establish the consistency of mathematics by establishing the consistency of a logistic axiomatization of mathematics such as *PM*. They hoped to establish the consistency of a logistic axiomatization by finitistic means. Here "finitistic" means that all the statements made in establishing the consistency of the axiom system would be justified by reference to uninterpreted formulae or finite sets of uninterpreted formulae of the axiom system. (If it were not for the fact that Formalists would use mathematical induction in their consistency proofs, I believe that their finitism is close to what I have called strict finitism.) But the nature of Formalistic metamathematics is not important for this critique of Formalism. What is of interest here is the fact that Formalists regard a logistic axiomatization of mathematics as mathematics. If they did not, they would not have tried to establish the consistency of mathematics by establishing the consistency of a logistic axiomatization. But in regard to mathematics being a logistic axiomatization of mathematics Formalists differ from Logicists. Formalists hold that an uninterpreted logistic axiomatization is mathematics while Logicists identify the interpreted axiom system with mathematics. (I should note however that Formalists such as Haskell B. Curry include the meaningful metalinguistic statements about a logistic axiomatization as part of mathematics. See Curry's "On the Definition and Nature of Mathematics" reprinted on pp. 152 to 156 of *BP*.) The identification of actual mathematics with an uninterpreted axiom system compounds the logicists' error of identifying mathematics with a logistic axiomatization. Once one accepted that mathematics could be properly presented as an uninterpreted axiom system, it is easy to drift into the position that mathematics is a game with symbols.

Of course, we can now take all of Wittgenstein's arguments against identifying mathematics with a logistic axiomatization and treat them as objections to Formalism. However, in IV-4 we find two objections which seem directed against the Formalists' additional error, – additional to the erors which they share with Logicism –, that mathematics is a game with symbols. In IV-4 Wittgenstein first asked how a Formalist distinguishes the game he calls mathematics from other games. The argumentive force of this question is to show that a Formalist will have to say that his game of deducing formulae is mathematics because these formulae represent mathematical propositions which are not uninterpreted formulae. Secondly, he pointed out that a philosopher who claims that mathematics is a game with symbols will not be able to talk properly about what mathematicians actually do. A Formalist would speak absurdly if he said that mathematicians pass out of mathematics when they cease to do formal deductions in an uninterpreted calculus. But a Formalist would have to hold this absurdity if he held that doing mathematics is really playing the game of constructing formal deductions. In III-35 Wittgenstein also observed that we do not regard mathematics as dealing with signs and he took this observation as showing that mathematics does not deal with signs. And he was right in so taking it as long as he wanted to restrict himself to philosophizing about what is now called mathematics. Or better: He overlooked only that small portion of current mathematics called metamathematics.

21. Because of its eccentricity, Wittgenstein's argument against a concern for consistency may not convince many that Formalism is in error. However, his arguments that a formal axiom system for mathematics is not mathematics seem to me to show that even if Formalists could establish the consistency of such axiom systems they would not have done much towards establishing the consistency of actual mathematics. When we pass from Wittgenstein's critique of Formalism to his critique of Intuitionism we have even less material. I read the following remark from Appendix II-18 as a mild criticism of Intuitionism.

> Finitism and behaviorism are quite similar trends. Both say, but surely, all we have here is . . . Both deny the existence of something, both with a view to escaping from a confusion.

I submit that here "finitism" refers to the Intuitionism of Brouwer and Heyting. A behaviorist says that surely we have only bodies, bodily processes, and bodily properties while an Intuitionist says that if we do talk about some objects in mathematics they surely do not exist independently of human thought. To avoid the confusion of thinking that there are minds and mental processes a behaviorist denies the existence of meaning to sentences which can be construed as indicating the existence of minds. Such a sentence would be: "Tom thought that Joan had something serious on her mind." An Intuitionist would deny the existence of meaning to sentences which suggest that there are mathematical objects existing independently of human thought. (I write the awkward "deny the existence of meaning to sentences" as an interpretation of Wittgenstein's "deny the existence of something".) For instance, the sentence "777 occurs in the decimal expansion of Pi" suggests that this expansion is laid out in some way outside of human thought and that it is the job of mathematicians to discover whether or not 777 is in this expansion. Hence, the Intuitionist denies the existence of meaning for this sentence about 777 in Pi. Wittgenstein differed from behaviorism and Intuitionism because his view was that we should try to understand these ways of speaking that suggest confusion. We should not try to eliminate them when we do psychology, mathematics, or speak in our ordinary language. So although Wittgenstein may not have been in disagreement with the mathematical ontology and epistemology of Intuitionism, I think that we can interpret this remark as showing that he disagreed with their way of doing philosophy. Intuitionists are Marxists in that they want to change language rather than understand it while Wittgenstein was Hegelian in that he wanted merely to understand language.

22. Of course, all the remarks in *RFM* which state or suggest strict finitism can be interpreted as criticism of a view that mathematical propositions are simply experimentally justified propositions. Strict finitism is an account of how experimental propositions are converted into mathematical ones. So if Wittgenstein was a strict finitist, we can conclude that he rejected a view that all mathematical propositions are simply well justified empirical truths. Also if strict finitism is an adequate account of even a small part of mathematics, we can conclude that not all mathematical propositions are simply well justified empirical propositions.

Wittgenstein also gave other arguments that mathematical propositions are not simply well justified empirical truths. In I-156 he cited as evidence of the non-empirical character of mathematics the fact that if we add 2 apples to 2 apples but somehow only have 3 apples altogether, we do not reject $2 + 2 = 4$ but say that somehow one must have disappeared. But he did not ignore the fact that empirical conditions are necessary for there to be mathematical truths although as we have seen he did deny that any empirical conditions were truth conditions for mathematical propositions. In II-64 to II-76 and also in V-14 and

V-15 he stressed that for there to be mathematics people must agree on calculations and each person must agree on his own calculations almost each time he does them. He also stressed that there must be at least enough order in nature so that we can carry out strict finitistic proofs. He went so far as to write in II-73:

But on the other hand I say again: 'Calculating is right – as it is done'. There *can* be no mistake of calculation in '12 × 12 = 144'. Why? This proposition has assumed a place among the rules.

But is '12 × 12 = 144' the assertion that it is natural to all men to work out 12 × 12 in such a way that the answer is 144?

23. Wittgenstein gave a negative answer to the question at the close of the passage just quoted. He did not conclude that mathematical propositions are anthropological propositions on how men infer and calculate. Nor did he conclude that mathematical propositions assert confusion in other people's calculations and our own proofs will not reign. The gist of his argument is in II-75. He pointed out that the conditions for speaking in a certain way, – here doing mathematics –, are not what is asserted by the propositions of that way of speaking. If we did hold that propositions assert the existence of conditions necessary for their use, then all propositions assert at least that there is not too much confusion for people to have language. But if all propositions assert that there is not too much confusion for people to have language then the proposition that there is not too much confusion to have language becomes a proposition which can be asserted by any proposition whatsoever. But since it is absurd to conclude that "There is not too much confusion for people to have language" can be asserted by any sentence we pick, it is absurd to conclude that propositions assert the existence of the conditions necessary for their use. (In this case, I think you could say that Wittgenstein gave what could be called a fairly standard type of philosophical argument.)

24. I do not think that Wittgenstein's criticisms of empirical interpretations are particularly original. However, I do believe that they are correct. The interesting aspect of Wittgenstein's thought on mathematical and empirical propositions is how closely

he connects them in his strict finitistic account of the creation of mathematical propositions.

25. We now come to an end of our examination of Wittgenstein's critique of other positions on both the philosophic and mathematical foundations of mathematics. Although it may be harsh to call these other positions madness, he has shown them to be defective. Our next task is to determine whether or not Wittgenstein's position is sanity in the midst of this madness.

V

ASSESSMENT OF *RFM*

> If I wished to find out what sort of thing arithmetic is, I should be very content to have investigated the case of finite cardinal arithmetic. For
> (a) this would lead me on to all the more complicated cases,
> (b) a finite cardinal arithmetic is not incomplete, it has no gaps which are then filled in by the rest of arithmetic.
> (*Blue Book*, p. 20 [Blackwell, 1958])

1. This opening quotation, although not from Wittgenstein's *RFM*, is, I believe, the kind of defense which Wittgenstein would give of the philosophy of mathematics developed in *RFM* against the most obvious objection to it. I believe that the most obvious objection to the strict finitistic philosophy developed in *RFM* is that it is at best a philosophy of elementary Euclidean plane geometry and what Wittgenstein called finite cardinal arithmetic, viz., addition, subtraction, multiplication, and division with finite numbers. In this chapter, we shall consider this and other objections to Wittgenstein's position and see whether or not he can meet some of these objections.

When I talk about criticizing Wittgenstein's philosophy in *RFM*, I should stress that I am talking about criticizing that philosophy of mathematics which I have called strict finitism. When I defend Wittgenstein against these criticisms, I will really be defending what I called at the end of Chapter III "open-ended strict finitism". On what basis shall I criticize strict finitism? Since Wittgenstein was not interested in providing mathematical foundations for mathematics and since a strict finitist would hold that each proof is sufficient foundation for the proposition it proves, there is no point in appraising strict finitism as providing mathematical foundations. However, for those who are concerned with mathematical foundations, strict finitism can be recommended as the best justified metamathematics to be used in consistency proofs, etc. But for my part, I shall appraise strict finitism on whether or not it is an adequate philosophy of mathematics or can be slightly altered so that it becomes one.

2. The criticisms of strict finitism as a philosophy of mathematics fall under two headings. The first criticism is that the strict finitists' paradigm of necessary truth for understanding mathematical necessity, viz., "The number of letters in 'Bismarck' is 8" is unacceptable. It is alleged to be unacceptable because it is said that mathematical truths cannot be construed as being this kind of proposition. Hence, the objection goes, even if strict finitists develop an appealing ontology and epistemology for necessary truths like their paradigm they still would have provided no philosophy of mathematics. (From now on I shall replace "construed as being of the same kind as the strict finitists' paradigm of mathematical necessity" with "falling under the strict finitists' paradigm".) The second criticism is that although strict finitists may be able to develop an appealing ontology and epistemology of a portion of mathematics by construing mathematical propositions as being of the same kind as "The number of letters in 'Bismarck' is 8", acceptance of "The number of letters in 'Bismarck' is 8" as a necessary truth requires an undesirable ontology and epistemology in general.

3. The criticism that mathematical propositions do not fall under the strict finitists' paradigm can have a strong and a weak sense. In the strong sense, it is the claim that no mathematical proposition falls under the strict finitists' paradigm. In the weak sense, it is the claim that some, actually many, mathematical propositions do not fall under the strict finitists' paradigm. In Chapter I, section 30 I pointed out that if a mathematical proposition fell under the strict finitists' paradigm, this meant that the paradigm proposition was sufficiently like the mathematical

proposition. From what was said in section 15 of Chapter IV it follows, then, that if a mathematical proposition M falls under a strict finitists' paradigm the mathematical truth expressed through M could just as well have been originally proved as a strict finitistic proposition. Hence, the strong sense of the above criticism is that if we had started out by doing strict finitistic mathematics we would not yet have done anything which we now call mathematics. The proof in Chapter II, section 26 of the strict finitistic version of 2 + 3 = 5, viz., "The counting together of 2 and 3 is counting 5", shows the absurdity of the strong sense. Indeed I suspect that the first mathematics that many people do is done in a strict finitistic way. (Here by "do mathematics" I mean "prove". I do not mean merely compute by means of rules.) So at least a portion of arithmetic can be done, i.e., rules of computation justified, by talking only about observable patterns and manipulations of these patterns to which arithmetic terms apply.

4. Simple geometric propositions such as "The shortest distance between two points is a straight line" can be established by strict finitistic means. To see this, consider Picture 4 below.

PICTURE 4

Stage 1: A B

Stage 2: A ——— B (Taking shortest route from A to B.)

Stage 3: A ——— B (Recognizing that the picture of taking the shortest route is a straight line.)

In Stage 1, A and B are perspicuous as points. If we would not call tiny dots points, there is nothing observable that we would call points. But it is proper to talk about observable points; we ask for the distance from one point to another. Stage 2 is a memorable picture of the operation of going from A to B by the shortest route. Stage 3, of course, looks the same as Stage 2 but it is a different stage because we are looking at it in a different way. In Stage 3 we judge the quality rather than the quantity of the line as we did in Stage 2. We can identify the line in Stage 3 as a straight line if it lies along a straight edge or if it is what people call straight. Once we identify the line as a straight line, Stage 3 can be surveyed so that we see that drawing a picture of the shortest distance between A and B is drawing a straight line between A and B. Note that this truth about points and lines was established without any reference to extensionless points or breathless lines. This is why I speak of strict finitism having an appealing ontology and epistemology. Another example of a geometric truth established by strict finitistic means is the proof of: "The square constructed on the diagonal of a given square is a square with twice the area of the given square" given by Socrates in the *Meno*. But, of course, the doctrine of recollection is not part of strict finitistic philosophy.

5. Hence, strict finitism as a philosophy of mathematics is superior to Logicism because strict finitists philosophize about at least a portion of mathematics since they philosophize about that portion which can be done by strict finitistic means. But the Logicists philosophize by talking about what is not and cannot be mathematics, viz., a logistic axiomatization.

6. We must now consider the weak sense of the claim that mathematical propositions do not fall under the strict finitists' paradigm. The weak sense is that MANY mathematical propositions do not fall under the strict finitists' paradigm. In the weak sense, the claim is true; and this, in my opinion is the only serious defect in the philosophy of mathematics developed in *RFM*. We have seen in Chapter I, section 20, 21, and 22, while examining his remarks on Cantor's and Dedekind's theorems, that as presently interpreted these theorems would not fall under the strict finitists' paradigm. But unfortunately we do not have to go to the mathematics of the real numbers to find mathematical propositions which do not fall under the strict finitists' paradigm. Mathematical induction cannot be accepted by a strict finitist. For a strict finitist a principle of induction would have to read as follows. (I shall write "provable by strict finitistic means" as SF provable.) For

any operations f and g, if it is SF provable that f (1) = g (1) and if it is SF provable that f (n–1) = g (n–1) then it is SF provable that f (n) = g (n); then for all natural numbers n, it is SF provable that f (n) = g (n). A strict finitist could not accept this principle of induction because he could not find a memorable picture of applying one operation which would also be a picture of applying any operation. In other words, the quantification over operations removes it from the realm of strict finitistic propositions. Also a strict finitist would never assume that it is SF provable that f (n–1) = g (n–1) for n–1 picked at random. He would not make such as assumption because he realizes that if n–1 is too large to get a perspicuous set of objects for applying n–1, he may not be able to have a strict finitistic proof that f (n–1) = g (n–1). So, since a strict finitist would not make an induction assumption, he could not use induction even if something led him to accept the principle.

7. However, propositions even more elementary than principles of mathematical induction fail to fall under the strict finitists' paradigm. Addition, multiplication, and division with many numbers or with very large numbers cannot be justified by strict finitistic means; such calculations can be done only in the decimal. Actually, I suspect that when computations get very large they can only be done by electronic computers without any possibility of justification by strict finitistic means or by pencil and paper computations in the decimal notation. To be sure, a strict finitist could use abacuses to get proof-patterns. But sooner or later the abacuses would get so huge that they could not justify computations done in the decimal notation. Of course, Wittgenstein was aware of the fact that strict finitistic proofs, i.e., proofs in the stroke system, cannot justify most computations in the decimal notation. He pointed this out in II-51, II-54, and II-57. But I do not find in *RFM* any account of how mathematical truths are justified when they are not justified by strict finitistic means. Maybe I have not found such an account because I have blinded myself to any conventionalism in Wittgenstein's philosophy of mathematics done within a system.

8. So we must conclude that strict finitism, and hence Witt-

genstein's philosophy of mathematics as well, are philosophies of only a little bit of what is now called mathematics. I suppose that strict finitism would be a philosophy of all of mathematics if the average IQ were forty points lower. But such a conjecture hardly puts strict finitism in a favorable light.

9. Is there any defense of strict finitism? There is no defense for strict finitists who hold that all mathematics should be done in a strict finitistic way since this cannot be done. Nor can there be any defense of a strict finitist who claims that we should eschew all mathematics which cannot be done by strict finitistic means. A philosopher ought not urge a return to barbarism. Similarly, philosophers of mind or philosophers of perception ought not to urge a return to the barbaric stage of talking only of bodies or sensations. Part of the point of Wittgenstein's claim in *PI*-124 that philosophy leaves everything as it is, is a rebuke of such primitivism.

10. However, the opening quotation of this chapter shows two mutually compatible ways of defending strict finitism as a philosophy of mathematics. Point (a) of that quotation suggests defending strict finitism by adopting what I called, at the end of Chapter III, open-ended strict finitism. An open-ended strict finitist accepts strict finitism as an adequate philosophy of only as much mathematics as can be done by strict finitistic means. However, he resolves to understand the remainder of mathematics by deviating as little as possible from strict finitistic philosophy.

In particular an open-ended strict finitist will resolve not to abandon the following beliefs (and I believe that I will be listing positions which Wittgenstein would be reluctant to abandon). We do not know what a proposition means as a mathematical proposition unless it is proved. A mathematical proof must show us what it is like for the proved proposition to be necessarily true. Mathematical truths must be proved but they do not have truth conditions. Mathematical truths are not determined as true by conventions. Without certain psychological and physical facts there would not be the mathematical truths which we now have because we would not have the proofs which we now have although these empirical facts are not truth conditions for mathematical propositions. Extant mathematics is sufficiently justified as long as each proposition is backed by a convincing proof; there is no need for a mathematical foundation for mathematics. Mathematical foundations are only more mathematics. Of course, an open-ended strict finitist will have to abandon the belief that all terms which appear in mathematical propositions have application to some observable paradigm for applying that term. In other words, an open-ended strict finitist cannot demand that all proofs be pictorial. Unfortunately, it is not at all clear how the analysis given of pictorial proofs will lead us on to give an analysis of the more complicated cases where the analysis of the more complicated cases must be such that it preserves the above strict finitistic beliefs. So a strict finitistic philosopher of mathematics is not refuted by the severe limitations of strict finitistic mathematics if he shifts to open-ended strict finitism. However, if he wants to develop a philosophy of all extant mathematics he has only a pious anti-platonistic wish without any clear program of how to fulfill it.

11. Still a strict finitist's wish is not completely unfulfilled since his philosophy does cover the arithmetic of (small) finite cardinal numbers. And it covers it well. Point (b) of the opening quotation suggests defending strict finitism by pointing out that it does well what it does do. I have, in effect, been arguing throughout this essay that strict finitism is a good philosophy of (small) finite cardinal arithmetic. So I shall not here defend it on that point. Here, I shall consider whether or not strict finitism is worthy of attention because it does a good job in a small area although it fails elsewhere. In his *Blue Book*, pp. 17 to 20, Wittgenstein claimed that we should be satisfied with understanding particular cases and that we should not try to attain a general account. On p. 44 of his *Blue Book* he pointed out that putting a few books in order is progress when we organize a library, even if we do not know how these books fit in with the other books. According to Wittgenstein, a similar situation exists in philosophy. Attaining philosophic clarity about one small area is philosophic progress even if we do not know how this philosophic solution will fit in with others. So clearly Wittgenstein would not doubt

that strict finitism is a satisfactory philosophy of the mathematics which it does cover simply because it covers only a small amount of mathematics. However, Wittgenstein's satisfaction with a partial account of mathematics is based on a premiss which is unjustified unless he has a philosophy of all of mathematics. He needs a premiss that there are different kind of mathematics. I.e., he needs a premiss that there is not one body of interdependent mathematical truths nor is there one ontology and epistemology of mathematics. The clause: "a finite cardinal arithmetic is not incomplete", from point (b) of the opening quotation shows that he held such a premiss. But such a premiss cannot be justified until one has a satisfactory philosophy of all of mathematics. Until we have philosophized about all of mathematics we cannot be certain that we will not have to revise some of what we have done before. So although strict finitism may be defended by pointing out that it does well what it does do, this defense is meager since a bit more philosophy of mathematics may reveal that we were mistaken in thinking that strict finitism was a satisfactory philosophy of (small) finite cardinal arithmetic.

12. As we have just seen, strict finitism as a philosophy is far from being well justified because of the severe limitations of strict finitistic mathematics. We now come to our final question. Is strict finitism unacceptable as a tentatively held philosophy of a part of mathematics because it has unacceptable philosophic implications outside the philosophy of mathematics? It would be dogmatic to reject strict finitism simply because acceptance of strict finitism requires acceptance of *synthetic a priori* truths. Exhibition of such truths as "The reverse of the letters of OVER is REVO" refute the dogma that there is only logical necessity. If one wants to reject strict finitism because it accepts non-logical necessity, he must attack the ontology and epistemology which form a strict finitist's explication of this kind of necessity.

13. I shall appraise only one aspect of the ontology and epistemology underlying strict finitism's explication of non-logical necessity. This is the ontological claim that observable objects have essences conjoined with the epistemological claim that we invent rather than discover what these essences are.

Let me expand a bit on strict finitism's doctrine of essence. Individual observable objects, and not kinds of objects, have essences. But individuals have essences as being one of a kind. Thus, if I write the Bismarck-token, Bismarck, it is of the essence of this token that it have eight letters. But this token must have eight letters only because if it did not it would not be a Bismarck-token. So before we can determine that a property P of an object is essential to it we must first identify it as one of a kind. Once we have identified it as one of a kind, then if we cannot conceive of the object as being of the kind without having property P we decide that P is essential to the individual object. We saw in Chapter II, section 10 that we do not determine that an object must have a property P to be of kind K by logical deduction from the concept of being of kind K. Often when we determine that object X of kind K must have property P to be of kind K, we are just then including having P as part of our concept of being of kind K. For instance, only after we determine that a particular OVER must have a REVO as its reverse to be an OVER, would we include having REVO as its reverse as part of our concept of being an OVER.

What if anything is wrong with this doctrine of essence? It certainly avoids the difficult epistemological problem of accounting for how we discover which properties of a member of a kind are essential to it so that it can be of that kind. But the claim that we invent the essence leads to the following kind of objection. Someone may object that this doctrine of essence is based on the belief that objects are what we take them to be. This objection correctly calls attention to the fact that if a strict finitistic type proof does not lead us to decide that an object X must have a property P to be of kind K, then P would not be essential for being a K. Thus if we did not prove that an OVER must have REVO as its reverse, having REVO as its reverse would not be essential for any figure to be an OVER. A person, in this case, could admit the possibility of there being OVERs whose reverse was not REVO. (In *RFM* V-42 to V-44, Wittgenstein gave examples of how people could have different concepts of what it is to be of a certain kind from those which we normally have.)

So a strict finitist is forced to admit that he believes that two groups of people could identify an object as being of a particular kind and yet correctly go on to develop mutually inconsistent concepts of what it is to be of that kind. For instance, a strict finitist would admit that two groups of people could identify a figure as an OVER. One group could then go on to say correctly that it must have REVO as its reverse to be an OVER. But the other group could say correctly that it need not have REVO as its reverse to be an OVER. Of course, we have a hard time seeing how the second group could say this since we belong to the first group.

What is wrong with saying that inconsistent concepts of OVER can both be correct? It may seem to entail that in and by themselves OVER-configurations are nothing definite; they are an inchoate something which can be taken as anything. However, all that a strict finitist means when he says that mutually inconsistent concepts of a kind of object may both be correct is that both are correctly formed, viz., based on proof. Since he claims that essences are invented, he can place the inconsistencies in language; not in things. And he will remind us that there is nothing odd about there being inconsistencies in peoples' concept of a thing because people do see things differently. So I do not think that strict finitism's doctrine of essence leads to the absurd metaphysics that things are nothing definite before we look at them in a certain way. In Chapter III of my dissertation I incorrectly attributed such a metaphysics to strict finitism and Wittgenstein.

Of course, I have not defended strict finitism against attacks on all of its philosophic implications; that task is beyond the scope of this essay. In regard to strict finitism, the task of this essay has been to bring it out of *RFM* so that it can be examined philosophically.

14. In conclusion, I submit that in *RFM* Wittgenstein was what I have called an open-ended strict finitist. We have seen the inadequacies of such a philosophy of mathematics. It is not conclusively refuted but it is not clear that it can be fully developed. So I cannot recommend Wittgenstein's philosophy of mathematics

as the correct one. Nor can I recommend is as one likely to be correct. The only reason I can give for trying to extend strict finitism to all of mathematics is that it is such a pleasing account of small finite cardinal aritmethic and simple geometry that if one tries to work out the belief that it can be extended he will come to understand how it can be extended.

BIBLIOGRAPHY

Anderson, Ross Alan, "Mathematics and the Language Game", *Review of Metaphysics*, 11 (March, 1958). Also, *BP*, pp. 481-489.

Benacerraf, P. and Putnam, H., editors, *Philosophy of Mathematics: Selected Readings* (Prentice Hall, 1964).

Berkeley, G., *Berkeley Essay, Principles, Dialogues*, edited by Mary Whiton Calkins (Charles Scribner's Sons, 1929).

Bernays, Paul, "Sur le Platonisme dans les Mathématiques", *L'Enseignment Mathématique*, 34 (1935). Translated as "On Platonism in Mathematics" and reprinted on pp. 274-286 of *BP*.

—, "Comments on Ludwig Wittgenstein's *Remarks on the Foundations of Mathematics*", *Ratio*, 2 (1959-60). Reprinted on pp. 510-528 of *BP*.

Black, Max, *The Nature of Mathematics* (Harcourt, 1933). I used the International Library of Psychology and Scientific Method edition of 1959.

Dummett, Michael, "Wittgenstein's Philosophy of Mathematics", *Philosophical Review*, 68 (1959). Reprinted on pp. 491-509 of *BP*.

Hilbert, D., "Über das Unendliche", Math. Annalen, 95 (1926). Translated as "On the Infinite" and reprinted on pp. 134-151 of *BP*.

Kant, I., *Prolegomena to Any Future Metaphysics* (Library of Liberal Arts edition).

Kielkopf, Charles, "An Examination of Ludwig Wittgenstein's *Remarks on the Foundations of Mathematics*". A dissertation (University Microfilms, Ann Arbor, 1962).

—, "Deduction and Intuitive Induction", *Journal of Philosophy and Phenomenological Research*, XXVI, No. 3 (March, 1966).

Kreisel, G., "Wittgenstein's Remarks on the Foundations of Mathematics", *British Journal for the Philosophy of Science*, IX, No. 34 (August, 1958).

Wittgenstein, H., *Tractatus Logico-Philosophicus* (Routledge & Kegan Paul, 1922).

—, *The Blue and Brown Books* (Basil, Blackwell, 1958).

—, *Philosophical Investigations* (Macmillan, 1953).

——, *Remarks on the Foundations of Mathematics*, edited by G. H. von Wright, R. Rhees, G. E. M. Anscombe and translated by G. E. M. Anscombe (Macmillan, 1956).

The above works are only the works from which I have quoted or which I have paraphrased a part. A more comprehensive bibliography is in *BP*; also there is one in *Ludwig Wittgenstein The Man and His Philosophy* by K. T. Fann (Dell Books, No. 5116). Several discussions of RFM have appeared since I completed this manuscript in October 1967.

INDEX

* This index makes no pretence of completeness. In most cases, only the most crucial uses of the term have been cited. The analytic table of contents may be of more help.

Bei Fragen zur Produktsicherheit wenden Sie sich bitte an:
If you have any questions regarding product safety,
please contact:

Walter de Gruyter GmbH
Genthiner Straße 13
10785 Berlin
productsafety@degruyterbrill.com